A Manager's Guide to Wireless Telecommunications

A Manager's Guide to Wireless Telecommunications

RON SCHNEIDERMAN

AMACOM

American Management Association

New York • Atlanta • Boston • Chicago • Kansas City • San Francisco • Washington, D.C.
Brussels • Mexico City • Tokyo • Toronto

384.5
S35m

This publication is designed to provide accurate and authoritative information in regard to the subject matter covered. It is sold with the understanding that the publisher is not engaged in rendering legal, accounting, or other professional service. If legal advice or other expert assistance is required, the services of a competent professional person should be sought.

Schneiderman, Ron.
   A manager's guide to wireless telecommunications / Ron Schneiderman
      p.   cm.
   Includes index.
   ISBN 0-8144-0449-9
   1. Wireless communication systems—United States.   2. Business enterprises—United States—Communication systems—Management.   I. Title.
   HE8678.S36   1999
      384.5—dc21   98–31108
                                                                              CIP

© 1999 Sci Tech Publishing, Inc.
All rights reserved.
Printed in the United States of America.

Printing number

10 9 8 7 6 5 4 3 2 1

JK

for Susan (again)

Contents

Prologue   ix

CHAPTER 1
An Introduction to Wireless Communications Services   1

CHAPTER 2
Fixed Wireless—Hard Connections   33

CHAPTER 3
Mobile Communications Satellites—
Roaming the World   53

CHAPTER 4
Wireless Data—More Than Just the Internet   77

CHAPTER 5
Mobile Computing—You Can Take It With You   89

CHAPTER 6
Wireless Telecom Regulations—Confusion and
Competition   103

CHAPTER 7
What's Next?—Smaller, Lighter, Multifunctional   121

Industry Sources   137

Glossary of Terms and Acronyms   147

Index   169

Prologue

For most business professionals, the thought of being at work means being at a desk in an office. That's changing, and rapidly, as business professionals have become more mobile and spend more time away from their offices.

One-third of the U.S. workforce, or about 43 million Americans, are now mobile, spending at least 20 percent of their time away from their primary workplace. These numbers are expected to increase as more companies, driven by simple economics and the demands of competition, push more of their workers out into the field and abolish fixed office space.

This trend certainly bodes well for vendors of wireless communications equipment or services, especially when you consider that twenty cents out of every dollar in revenue generated by telecommunications service providers comes from wireless communications. IBM Corp. believes that by the year 2008, that figure will climb to eighty cents.

Most businesspeople already carry a cellular phone or a pager (sometimes both) and may regularly use a laptop computer and a personal information device of some type. The technology in these devices is advancing so fast that the functionality that used to be available in separate products can now be integrated into a single product, even as products become smaller and lighter. Multifunctional products that are part cellular phone (voice), part handheld personal computer (data), and part portable navigation system (position-location) are fast emerging.

As business users account for a growing share of more broad-based wireless service users, wireless service providers will develop more new products and services for the business/professional market. Enterprise wireless communications—the wireless office, in-building cellular, the wireless PBX—will become critical elements in boosting the productivity of every business, including small offices and even home offices.

Wireless technology allows us to communicate not only more efficiently but also more creatively. When Boeing invited several VIPs and a few members of the press to take the inaugural flight of its new 777, a photographer from the *Seattle Post-Intelligencer*, the local morning daily, took along a digital camera. He shot several pictures during the flight, then dumped them into his laptop computer. He selected four pictures and then, using the aircraft's satellite-based seatback phone, downloaded the photos to his office. The pictures were on the press before the plane landed.

While the Internet has dramatically changed the way companies conduct their business, corporate intranets are becoming increasingly pervasive, so much so that an estimated 33 percent of large U.S. corporations are expected to provide their field service and sales personnel with wireless intranet access by the year 2000.

How can CEOs and other key business managers respond intelligently to the few people in their companies who make recommendations on how to spend millions of dollars to install, operate, and upgrade their company's telecommunications network?

Your information systems (IS) or information technology (IT) department is probably throwing a lot of mobile communications options at you, and it's going to cost a lot of money. You may not be sure what they're talking about. This book is designed to explain many of the growing options available to you and your business, as well as some of the emerging and developing wireless communications products and services that may well serve you and your business.

A Manager's Guide to Wireless Telecommunications

CHAPTER 1

An Introduction to Wireless Communications Services

Twenty years ago, even the most astute market analysts could not predict the revolution that resulted from the development of the personal computer. The wireless communications revolution is an even more recent phenomenon as businesses, large and small, are consuming these new and emerging telecommunication services at an ever-increasing rate. Wireless services have become nearly ubiquitous, with 60 percent of businesses currently subscribing to cellular/personal communications services. The question now is, what will it take to make wireless products as pervasive for business uses as the telephone or fax machine? And where do you and your business fit into this picture?

Wireless systems fall into two groups, depending on their range of operation:

1. *Wide area systems* are targeted at highly mobile users whose primary concern is the ability to communicate at any time from any location.

2. *Local area systems,* because they are more of a convenience, usually replace local wireline service.

A comparison of these two groups of wireless systems is found in Figure 1-1.

Figure 1-1. Wide- and local-area wireless communications.

	Wide-Area (High Mobility)	Local-Area (Low Mobility)
Voice Communications	■ Cellular Telephones ■ Satellite Systems	■ Cordless Telephones ■ Wireless Local Loop Systems ■ Wireless PBXs
Data Communications	■ Cellular Telephones ■ Pagers ■ Specialized Packet Data Systems (RAM, ARDIS, CDPD) ■ Satellite Systems	■ Cordless Telephones ■ Wireless LANs

There are several wireless services to choose from that can meet just about any business requirement. First, let's briefly define the origins of today's cellular services and define some important terms.

A Brief History of Cellular Communications

Cellular communications service started in the United States in Chicago in October 1983. Cellular service developed as an offspring of earlier work (most of it done during and shortly after World War II) and research undertaken by some of the biggest names in the radio industry, including Motorola, AT&T, and others.

AT&T recognized the commercial potential for mobile communications and, by 1946, the telecommunications giant had created Improved Mobile Telephone Service (IMTS), the first mobile radio system to connect with the public telephone network. AT&T quickly won approval from the Federal Communications Commission (FCC) to operate the first commercial public radio service in St. Louis. This service soon spread to twenty-five cities using high-power transmitters with a radius of about fifty miles each.

The next step was to figure out a way to extend service beyond this fifty-mile radius of coverage. Incredibly, it took twenty years to develop the technology that would enable mobile radio subscribers to "hand off" calls from one cell to another cell. (A "cell" is a geographic area that is served by a single low-powered transmitter/receiver. A cellular service is made up of multiple or overlapping cells.) A call from a mobile phone is essentially switched from one transmitter to another as the subscriber travels across a region or across the country.

By 1973, Motorola had demonstrated a mobile phone that worked with the AT&T-developed network. In 1977, the FCC authorized two experimental licenses—one to AT&T in Chicago, the other to Motorola and American Radio Telephone Service, Inc. in the Baltimore/Washington, D.C., area. In early 1981, the FCC announced that it would approve two licenses in each market—a non-wireline company (which became known as the "A" side carrier), and a wireline company (the "B" side carrier).

Realizing that the "competitive hearings" process that it had set up to grant licenses to cellular operators was too slow and too costly, the FCC announced in October 1983 that lotteries would be used to award licenses in all markets below the top-thirty systems. By 1984, Washington, D.C., had two competing cellular providers. By 1990, licenses had been issued for at least one system in every market in the United States. Cellular is now available virtually worldwide. By the end of 1997, there were more than fifty million cellular subscribers in the United States alone.

How Do Cellular Communications Work?

In cellular communications, each cell is served by its own radio telephone and control equipment. Each cell is allocated a set of voice channels and a control channel with adjacent cells assigned different channels to avoid interference. The control channel transmits data to and from the mobile or portable units.

These control data tell the mobile/portable unit that a call is coming from the mobile telephone switching office (MTSO) or, conversely, tells the controller that the mobile/portable unit is placing a call. The MTSO also uses the control channel to tell the mobile/portable unit which voice channel has been assigned to the call. The 25 megahertz (MHz) of frequency spectrum assigned to each cellular system currently consists of 395 voice channels and 21 control channels.

Low-power transmitters are an inherent characteristic of cellular radio systems. As a cellular system matures, the effective radiated power of the cell site transmitters is reduced so channels can be reused at close intervals, thereby increasing subscriber capacity.

However, concern arose that so many cellular subscribers and cell sites were operating in the United States, and that the basic analog cellular system, known in the United States as AMPS (Advanced Mobile Phone Service), was approaching system capacity in many of the larger cellular markets. The industry therefore took steps to meet the future demands for service by developing new standards using various digital modulation techniques that increase system capacity. Today, digital cellular technologies are gaining acceptance worldwide, particularly in Europe and the United States, along with a wide range of new services.

Why Digital?

Digital technology offers several advantages over analog technology, among them:

- Improved sound quality
- Longer battery life
- A higher level of privacy and security

Digital wireless technologies also provide a platform for new features, including short text messaging and caller ID. Digital also makes more efficient use of the frequency spectrum so that it can carry more traffic for less cost while transmitting voice, data, and other wireless services. Presumably, these lower costs will be passed along to the customer.

Criteria for Evaluating a Digital Service

Four very important features that have helped make digital wireless service attractive to growing legions of users—in addition to the obvious advantages of mobility or portability—are what have become known in the industry as the Four C's: clarity, cost, coverage, and capacity. All of these are extremely important in selecting the right wireless phone and service.

1. *Clarity.* While generalizations can be made about the relative quality of the major digital systems, field trials by service providers and surveys by analysts indicate that voice clarity can vary noticeably from one digital technology and manufacturer's product to another.

2. *Cost.* You don't have to look at too many advertisements to realize that pricing plans vary all over the lot. In fact, the introduction and promotion of new services, such as PCS, have driven cellular service prices down and have even made wireless services more competitive with basic wireline telephone services.

3. *Coverage.* In terms of coverage, the selection of a service provider usually requires a little more homework by the subscriber because of the differences in the "footprints," or coverage areas, held by regional and national carriers and how they impact cellular and even cordless phone service. Where it starts to get complicated is with the introduction of dual-mode (i.e., analog and digital) and dual-band (i.e., phones that operate at two frequencies) handsets, and how these devices work for the mobile professional.

Dual mode phones are necessary (at least in the United States) because it is going to take some time before every region of the country has digital service. This means that if you have a dual-mode phone, you can use it anywhere in the country. The phone will seamlessly switch from digital to analog and back, depending on what service (or technology) is available in the area you are in at the moment.

4. *Capacity.* Capacity remains a vague issue at the moment as digital service providers are making claims of capacity gains of ten times that of analog systems. For the most part, however, the real-world capacity gains of digital service are still unknown and have not been fully taken advantage of in terms of creating meaningful new features and services. A highly controversial technical issue related to capacity is the ongoing battle among equipment suppliers and service providers over which "air interface" standard offers the best service. There are three primary competing standards:

1. *Time division multiple access* (TDMA) uses time-sharing to accommodate multiple users on each communication channel to increase system capacity. With TDMA, each user is allocated a fixed interval for transmission and then the next signal is allocated the same interval. The process is repeated continuously, with the interval for any particular signal being providing enough frequency enough to accomplish voice communications. TDMA proponents claim up to three channels of capacity over the single channel available from analog cellular systems.

2. *Code division multiple access* (CDMA) is a different system. Whereas traditional systems transmit a single strong signal on a narrow band, CDMA works in reverse, sending a weak but broadband signal. A unique code "spreads" the signal across a wide area of the spectrum (a technique called "spread spectrum"), and the receiver uses the same code to recover the signal from the noise. By using different codes, a number of different channels (ten or more, according to most CDMA carriers) can simultaneously share the same spectrum without interfering with one another.

3. *Global System for Mobile Communications* (GSM), which was developed in Europe, is a TDMA-based system whereby users share a frequency band. In GSM, each user's speech is stored, compressed, and transmitted as a quick packet of information. The time slots can be controlled and therefore distinguished from the others. (Hence, the term "time division.") The packet is then decompressed at the receiving end of the transmission. GSM, like the other digital systems, also has extended capacity—up to eight users share a given channel. (GSM is the standard digital cellular system throughout most of Europe and several other areas of the world and has been adopted by several cellular PCS carriers in the United States.)

Cellular Versus PCS: What's the Difference?

Whatever system dominates (i.e., TDMA, CDMA, or GSM), digital technology is gaining. An estimated 25 percent of the world's cellular subscribers are currently using digital technologies. This figure is expected to grow to at least 45 percent by 2000. (Cellular and PCS subscribers should top 207 million worldwide in 1998.)

In fact, a new service, known as personal communications services (PCS), is already helping to promote all-digital (i.e., TDMA, CDMA, or GSM, among at least four other technologies) service.

Admittedly difficult to differentiate from cellular service, PCS has been dubbed "plagiarized cellular service" and "cellu-

lar at a different frequency." In fact, in terms of their functionality, cellular (particularly digital cellular) and PCS are very
similar and many cellular carriers now offer PCS.

AT&T Wireless, in an effort to downplay the differences
between cellular and PCS, launched its digital TDMA cellular
service across the country under the name AT&T Digital PCS.
This confused most consumers, but it could be a temporary
condition because AT&T eventually plans to integrate this service with its TDMA service at 1,900 MHz in areas where AT&T
holds PCS licenses. The two services will operate seamlessly
at that point, and AT&T will make no distinction in the brand-
ing of its digital cellular or PCS services.

The key difference between cellular and PCS lies in the
spectrum allocation:

- Cellular service operates at 800 MHz.

- PCS operates at 2 GHz.

The two services will become increasingly blurred as the
services are bundled and integrated in the future.

PCS grew out of a report published in Britain in 1989
called "Phones on the Move: Personal Communications in the
1990s." Just seven pages long, the report essentially jump-
started the telecom community in the United Kingdom into
offering mobile telecommunications. The report stated,
"More and more U.K. business is coming to rely on mobile
communications, and government has acted as an enabler,
making sure they get the services they need. . . ."

The British government responded to the study by quickly
licensing four companies to provide what were called telepoint or CT-2 (cordless telephone, second generation) services. Using essentially cordless pay phones, subscribers
could originate, but not receive, short-range phone calls in
public areas equipped with telepoint base stations, such as
train stations, airports, and busy shopping centers.

Telepoint made sense for London, which had very few pay
phones. Unfortunately, high subscriber costs ($200 for a

handset, plus a $60 service connection fee and a monthly service charge of $15) didn't sell well in a weak British economy. Also, the licensees were using incompatible technologies. As a result, the service never really caught on, leaving the British government to issue additional licenses to other telecom ventures (some of which had American partners) to develop what are generally known in Europe as personal communications networks (PCNs).

Lessons Learned From Telepoint

Despite all of its problems, the introduction of telepoint in the U.K. provided a wake-up call for the rest of the telecommunications world. In the United States, the FCC responded by issuing more than 200 experimental licenses for PCS trials. Cellular telephone carriers, cable television system operators, independent telecom system service providers, and telecom equipment manufacturers began spending millions of dollars to develop new services and products in anticipation of the national deployment of PCS.

A study commissioned by the Cellular Telecommunications Industry Association (CTIA), a U.S. cellular industry trade association, helped determine what the British had done wrong. One problem was the high cost of building three separate PCN infrastructures (costing about $1 billion each). Another key factor was the way the British government licensed radio frequencies for the PCNs. It simply gave the spectrum away. U.K. cellular operators complained not only about the giveaway, which resulted in new competition for the cellular carriers, but also about the fact that the PCNs were being assigned more spectrum (50 MHz each) than the cellular operators were licensed to use (30 MHz).

PCS Deployment in the United States

The FCC's expectations that PCS will "compete with existing cellular and private advanced mobile communications services, thereby yielding lower prices for existing users of those services," has pretty much taken place. In many parts of the United States, PCS has already led to the development of a wide range of services and user devices such as smaller, lighter, multifunction portable phones, multichannel cordless phones, and wireless PBXs (private branch exchanges). PCS already competes with both cellular and paging services, but the combination of PCS licensees and cellular and paging carriers responding to the new competition with more advertising and promotion has significantly boosted the number of subscribers for all of these personal communications services.

With authorization to auction PCS licenses from the Omnibus Budget Reconciliation Act of 1993, the FCC decided to adopt a combination of Major Trading Areas (MTAs) and Basic Trading Areas (BTAs) to promote the rapid deployment of PCS. The MTAs and BTAs were designed by Rand McNally and were based on the results of studies covering population distribution, economic demographics, and even newspaper circulation. The FCC also wanted PCS to facilitate regional and nationwide roaming and allow licensees to tailor their systems to the natural geographic dimensions of PCS markets, as well as technical standards and the cost of coordinating interference between PCS licensees.

Initially, six companies won licenses to provide nationwide, narrowband PCS services, mainly for two-way paging services. Another nine companies won licenses to provide regional, narrowband PCS in a second round of auctions.

Broadband PCS, on the other hand, is much more like cellular in that it offers wireless voice and data transmission over both local and wide areas using low-power, lightweight portable phones and handheld computers.

Licensed and Unlicensed PCS Spectrum

There are now 2,074 PCS licensees in 51 MTAs and 493 BTAs in the United States. Several of these licensees were awarded "Pioneer's Preference" status by the FCC, which means that they received their license free by developing new communications services and technologies that would improve existing services.

There is also an unlicensed portion of PCS spectrum outside the narrowband and broadband applications that is designed for the operation of short-distance wireless voice and data devices. These are regulated by the FCC under the so-called industrial/scientific/medical (ISM) bands and include wireless local area networks and wireless PBXs.

The expansion of unlicensed PCS is expected to lead to a big jump in in-building wireless communications equipment installations. Spending on PCS and other in-building telephone equipment totaled $99 million in 1996 and is projected to climb rapidly to $935 million in 2000, primarily on the strength of the growing unlicensed PCS market.

Other Wireless Services and Products

SMR and ESMR Networks

Specialized Mobile Radio (SMR) or Enhanced SMR (ESMR) is a two-way radio dispatch service established by the FCC in 1979 for the public safety, construction, and transportation (including taxi) industries. SMR services include voice, data broadcast, and mobile telephone service, but it currently has limited roaming capabilities.

An SMR subscriber can interconnect with the public telephone network much like a cellular subscriber. SMR systems use one large transmitter to cover a wide geographic area. This limits the number of users because only one user can talk on one frequency at a time. To complicate the problem, SMR operators have fewer frequencies than cellular service providers, and there have been more operators in each market. However, because most dispatch messages are short, the system has been able to handle a normal day's dispatch voice traffic.

There are two distinct types of SMRs—conventional and trunked systems.

1. A conventional system has only one channel. That means that if someone is already using a "conventional" channel, you can't use that channel until it is available.

2. A trunked system, in contrast, can accommodate several users at one time by combining channels and automatically searching for an open channel. Most current SMRs are trunked systems.

Although SMRs are primarily used for voice communications, systems are being developed for data and fax services. Like cellular, SMRs are changing over from analog to digital systems enabling the development of new features and services such as two-way acknowledgment paging and inventory tracking, credit card authorization, automatic vehicle location, flcct management, remote database access, and voice mail.

Motorola has designed a wireless handset called the I1000 (see Figure 1-2) in response to another group's needs—namely, business professionals. It accommodates their desire for more convenient features in a smaller, lighter communications device and integrates multiple communications services into a single device, including a digital cellular phone, two-way radio, and alphanumeric pager. Features include a speakerphone and caller ID.

The Nextel Network

In 1987, Nextel Communications, Inc. (formerly Fleet Call) began acquiring spectrum in six of the largest markets in the United States—Chicago, San Francisco, Los Angeles, New York, Dallas, and Houston. In 1990, Nextel asked the FCC for permission to build an enhanced SMR system in those six markets. Its ESMR system would require lower power and would feature multiple transmitters, enabling the same frequencies to be reused at longer distances. The result was a cellular-like system that allowed Nextel not only to extend its operations into new business applications but also to begin selling its services to some consumers. Nextel's first digitally based ESMR came online in Los Angeles in early 1991.

Figure 1-2. Motorola's I1000 wireless handset.

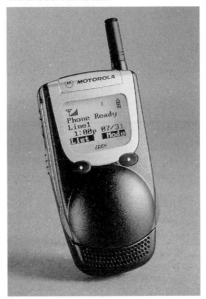

Nextel, with the help of a $1.5 billion investment from Craig McCaw shortly after he sold McCaw Cellular to AT&T in 1995, has acquired several additional SMRs throughout the country and is expected to continue to grow its network into a national system over the next several years. Unlike cellular systems, there are no roaming charges when using Nextel's service. Nextel charges the same per-minute fee to users who travel as when they are in their home base, plus a 15-cent, flat-rate long-distance charge.

Now based in McLean, Virginia, Nextel has dominated the SMR market to the extent that it had an estimated 300,000 "interconnected units" at the end of 1997, compared to its closest competitor, Pittencrieff Communications, Inc., with 36,800 interconnected units. Industrial Communications & Electronics, Inc. is a distant third with 11,000 units.

Nextel's business strategy is to tie different companies with common business interests together into virtual business networks. The new groups will be marketed under such names as the Builders Network, Transportation Network, and the Real Estate Network, with potentially hundreds of companies signing on in several major market regions. Plans also call for the introduction of a data service for business users.

Another ESMR company, Geotek, based in Montvale, New Jersey, has focused on small businesses, providing mobile voice and data dispatch services along with automatic vehicle location through a vehicle-mounted workstation.

Paging Networks

Paging continues to be the most reliable, cost-effective, and therefore the most popular form of wireless communications in use today. More than 40 million people subscribe to a one-way paging service worldwide, and that number could easily top 50 million by the year 2000. While advanced messaging is likely to continue to grow over the next few years, simple numeric paging remains an enormously popular service.

Price is the key to paging's success, and it just keeps dropping. A study by The Strategis Group has tracked the continued decline in average monthly revenue per unit from $10.51 in 1995 to $9.77 in 1996. However, the increasing popularity of alphanumeric paging and the development of advanced messaging devices and services (e.g., voice, two-way acknowledgment paging, and the growth of e-mail and Internet services)—as well as the stabilization of reseller rates—should have a strong positive impact on the average monthly revenue of paging and other messaging services. Some paging services also offer nationwide coverage, an important feature for traveling road warriors.

The more recently introduced two-way pagers have been slow to gain favor, and that isn't likely to change until service is improved and these devices are priced more competitively with other messaging services. In fact, when they were introduced in 1995, two-way pagers were priced at about the same level as many cellular phones.

Two-Way Messaging Trends

SkyTel and Wireless Access have created a second-generation two-way messaging device that can initiate messages directly from the pager. SkyTel 2-Way subscribers can also send and receive messages from Hewlett-Packard palmtop PCs and handheld organizers. Mtel, another paging service, also offers several two-way paging services.

Also, many of the new cellular and PCS phones feature built-in short text messaging capability, making them a more competitive choice over even the most advanced pagers. Since market research strongly indicates that most pager users would like to be able to reply, it is a fair bet that paging carriers will respond with reliable, if not national, two-way services.

Several new two-way pagers are currently available, including the Motorola PageWriter 2000 shown in Figure 1-3. In addition to interactive paging, the two-way model offers a variety of Internet-based services and general news and information features, such as weather forecasts, sports scores, traffic reports, movie schedules, and driving directions. All these services can be accessed wirelessly at predetermined intervals or on demand, enabling pager users to customize data to meet their personal needs.

Part of the problem for designers of paging devices is to come up with an advanced device with all the bells and whistles that is no larger than currently available pagers. How will paging subscribers initiate messages on such a small device? Do users prefer miniature keyboards or a pen-based touchscreen?

Some of the newer two-way paging devices will be able to initiate messages and increas-

Figure 1-3. Motorola's new two-way pager, the PageWriter 2000

ingly will support messaging with laptop and handheld personal computers (HPCs), as well as personal digital assistants (PDAs).

One supplier of two-way pagers, Ontario-based Research in Motion, offers a wireless PC card for Mobitex (the open, nonproprietary wireless data protocol developed by Swedish Telecom and Ericsson) in the United States. This product brings wireless connectivity to portable computing devices using the RAM Mobile Data network.

E-Mail and the Internet

Clearly, e-mail and the Internet are critical to the growing success of high-end, two-way, advanced messaging devices, including pagers. News items, sports scores, stock quotes, weather forecasts, and other basic services are gaining in popularity for these highly portable devices. Motorola, for example, has teamed up with ESPN Enterprises, an all-sports cable TV network, to provide ESPNet, a pager service that offers sports news and commentary. Panasonic held off introducing its first pagers until its market research indicated that pager users were willing to pay for information as well as a personal messaging service. (Unlike most pagers, Panasonic uses the FM subcarrier radio broadcast data system to send and receive information, the system already used by FM broadcasters as a data transmission channel.)

A more recent feature, introduced by Lucent Technologies, is the ability to receive turn-by-turn driving directions on a pager.

Digital voice paging, on the other hand, is relatively new. Subscribers can dial an 800 number and leave a voice message, which is then transmitted to the pager. Voice pagers can store and play back up to four minutes of messages.

A new type of pager that is expected to gain popularity operates like a portable answering machine. This pocket-sized unit allows messages to be listened to immediately or stored and heard at the user's convenience. Motorola's version is shown in Figure 1-4.

Many of the newer pager-based voice mail models will use a narrowband PCS network, which the FCC has explicitly designated for two-way paging applications. Several paging services have already positioned themselves to take advantage of these new frequencies to offer regional and nationwide two-way service.

Figure 1-4. Motorola's new pocket-size pager, which operates like a portable answering machine.

If real portability is important, Seiko Telecommunications and Sharp Electronics have developed digital watch/pagers that use the FM subcarrier network to provide personal messaging, news, sports, weather reports, and traffic information. Motorola and Timex introduced a similar product in early 1998. Texas Instruments has also entered the advanced messaging market with its Advanta Pro. About the size of a PDA, this device offers a QWERTY keyboard with a six-line, 32-character backlit display for viewing alphanumeric messages. It also functions as a calculator and electronic organizer.

Cordless Phones

Cordless phones continue to gain in popularity, and there are now more than 100 million of these phones in use in the United States, with something like 18 million new cordless phones sold in the United States every year.

However, despite these impressive numbers, the 900 MHz cordless models, which offer much longer range, significantly more security, and more reliability than the more popular 46–49 MHz models, account for only about 10 percent of the market. The primary reason is cost. The 900 MHz versions are priced from just under $100 to more than $200 at retail, or about twice the more popular models.

Nevertheless, the high-end models are expected to claim half the market by the end of the year 2000 as prices drop and consumers and business users opt for an upgraded product.

Wireless LANs

Wireless local area networks (WLANs) are essentially networks that allow the transmission of data and the ability to share resources, such as printers, without the need to physically connect each node, or computer, with wires. The acceptance of WLANs has been slow, mainly because of cost and the lack of a single technical standard that would allow users to purchase WLAN hardware and software from virtually any suppliers and know that it would be interoperable with anyone else's WLAN products.

The situation has changed because WLANs have dropped in price. In addition, an international committee of communications and computer specialists appointed to the task has formally adopted a technical standard for WLANs. Almost seven years in the making, the standard, which was approved by the Institute of Electrical and Electronics Engineers (IEEE) Standards Activity Board and is known as IEEE 802.11, covers just about everything—from wire augmentation and ad hoc networking to factory automation.

Wireless PBX

Wireless private branch exchange (WPBX) is another burgeoning growth area. As users' mobility needs grow within organizations, demand for wireless communications will likely increase in the future.

Most WPBX systems provide essentially the same services as any PBX, but they also give users remote access to their organization's phone system. WPBX systems, for example, include a wireless handset that is programmed to ring simultaneously with the desk phone. The user can answer either phone and can even switch between the phones during the call.

According to at least one market projection, one out of five or six office phones will likely be wireless by the turn of the century. By 2002, between 15 and 20 percent of the new PBX shipments are expected to be wireless.

As a result, manufacturers are introducing wireless products that can work with the 65 million wired business phones in use today.

Wireless Local Loop

Wireless local loop (WLL) is essentially the wireless version of the existing wireline local loop, which is the actual physical connection between your telephone and the central telephone switching station.

WLL is a potentially huge market, particularly in underdeveloped and developing countries where the telecommunications infrastructure is not nearly as mature or built out as it is in more developed regions of the world. Wireless systems can often be installed in far less time and at lower cost than traditional wired systems.

In North America, Western Europe, and Asia, many service providers are beginning to offer fixed wireless services in office and factory environments. Most of these services are linked to cellular networks when users leave their closed campus setting. Some companies are even offering supercordless phones that can be used without incurring airtime charges at home and then convert to the public cellular network when out of range.

Mobile Satellite Service

Mobile satellite service (MSS) is a more recent development in wireless communications. Several international companies—and the list is growing—are investing in the development of constellations of low-earth orbit (LEO) or medium-earth orbit (MEO) satellites. They will supplement the more heavily used and more expensive geostationary (GEO) satellites that currently orbit the earth at altitudes of 22,300 miles.

One noticeable difference in these systems is that because LEOs operate from a much lower orbit than the GEOs—normally about 455 miles for LEOs and 750 miles for MEOs—subscribers will not have to deal with the annoying quarter-second delay often encountered when talking on a GEO system.

Basically, there are two types of LEOs: the so-called Big LEOs, which provide portable and mobile voice, data, fax, and paging services on a global basis, and Little LEOs, which provide just about everything but voice service.

So far, more than $60 billion has been invested in the development of these systems, and several new, much more ambitious mobile communications satellite systems are on the drawing boards.

Essentially, there are two markets for these systems:

1. International business travelers, who prefer to have the same kind of mobility when traveling abroad that they have when using their cellular phones at or near their home base of operation.

2. Underdeveloped or developing countries where telecommunications services are not quite up to par with the rest of the world. In these areas, business subscribers are likely to use MSS as part of their fixed telecom network. Consumers in these regions will have access to MSS not only through the use of portable and mobile phones (although the service will be more expensive than cellular), but also through strategically placed MSS-based kiosks or "pay phone" stations.

Radio Frequency ID Service

Radio frequency identification (RFID) is another fast-growing wireless service that is finding wide application in security, access control, transportation, assembly-line management, and product and animal tagging. In fact, just about anything that needs to be identified is a candidate for RFID. For example:

- The U.S. Postal Service is working with Lockheed Martin Corp. and Symbol Technologies to provide 300,000 barcode scanning handheld computers to support a network that will enable mail carriers to confirm delivery of express, priority, certified, and registered mail.

- Savi Technologies, a Raytheon TI Systems subsidiary, is supplying RFID tags to the Pentagon to locate, monitor, and track supplies anywhere in the world. Radio tags will identify the contents of containers and stored goods, and the handheld computers, known as interrogators, will "read" the location and status information off the tags.

- Amtech Corp.'s Transportation Systems Group has supplied 78,000 tags and 98 readers for TraansGuide, a traffic-management system in San Antonio developed by the Texas Department of Transportation. The readers are positioned at fifty-three strategic monitoring sites throughout the San Antonio area. Tagged vehicles act as traffic probes to provide intelligent information tracking, measuring actual travel times and average speed between readers.

- Another popular application of RFID is electronic toll collection, such as the EZ-Pass system used for automatic vehicle identification along highways and for bridges and tunnels in the metropolitan New York City/northern New Jersey area.

RFID networks have certain advantages over other technologies. For one thing, they do not require a direct line of sight between the reader and the tag. Also, the tags are reusable. Most RFID tags can be specifically programmed with critical information, such as the number or description of the item being tagged. And they can be reprogrammed for a different item.

Some RFID tags can operate at distances from a few feet to several yards. Others can operate from 30 to 300 feet.

There are three types of RFID tags—active, passive, and transponder:

1. *Active* tags have their own internal power source, usually a battery. They can be designed to transmit at some preset interval or power up when being polled from a reader.

2. *Passive* tags have no internal power, although they may have a small internal battery to retain information in memory. However, passive tags receive their power through their antenna from the reader's interrogation signals.

3. *Transponders* are also passive in that they receive energy from the reader and use some of that energy to power up their circuitry and return a signal to the reader.

There are several other wireless applications that are new enough to business and industry and important enough to require a more extensive explanation in terms of how they meet business users' needs, how they work, and how they are being used today. In addition to WLANs, WLL, and WPBX systems (which were briefly described in this chapter and will be discussed in more detail in Chapter 2), there are also wireless data systems (see Chapter 4) and mobile computing applications (see Chapter 5).

Market Drivers and Obstacles

Wireless Service Pricing

Price is a significant driver in the wireless communications market, just as it is in any other market. The introduction of new wireless communications technologies and products, along with the very rapid growth of this market, has led to a wireless pricing environment that is both very dynamic and highly unsettled. As competition increases, wireless carriers will likely be under tremendous pressure to reduce prices to capture and retain customers.

Cellular carriers have already surrendered their long-held duopoly status (two carriers in each market) and are forced to compete with PCS by lowering service pricing. A study by The Strategis Group found that over a three-year period (1995–1997), service prices fell 15 percent to 30 percent, depending on the level of monthly use. Among individual markets, there was considerable variance in wireless pricing given the economic and competitive differences of these markets. To challenge the cellular carrier's incumbency, the study found that many PCS service providers have entered the marketplace at prices averaging 16 percent below cellular carriers. Cellular carriers responded by lowering their prices to better compete with emerging PCS services. With lower prices, broader coverage and—in many cases—more functionality, the cellular camp has been able to compete effectively with the newer PCS.

In a somewhat similar fashion, Japan's version of PCS, called Personal Handyphone System, or PHS, has suffered in the marketplace as Japan's cellular carriers responded with lower prices and other incentives—so much so, in fact, that several industry analysts wonder how long PHS can last as a viable service. Italy had a similar experience when Telecom Italia Mobile began offering a supercordless phone service called Digital European Cordless Telecommunications, or DECT. Regional cellular carriers in Italy managed to slow the growth of DECT service simply by cutting their service prices.

Pricing of paging services has declined since as early as 1992. In other markets, such as wireless local area networks, growth has been slowed by two key factors—the lack of an industry technical standard (a problem that has been solved with the adoption of the IEEE 802.11 standard) and the perceived high price of WLAN (like most new and highly competitive services, prices are dropping all the time).

Similarly, mobile communications satellite services will be very expensive at the start (probably $3 a minute to use the Iridium LEO satellite system), but prices will begin to drop when other services are launched and come into widespread service.

The Health Risks of Wireless Products

Several articles have been published in recent years about the potential health dangers of using wireless products.

In fact, all electrical devices, including TV sets and hair dryers, generate electromagnetic fields. The electromagnetic energy produced by wireless telephones in "non-ionizing." There is currently no known mechanism by which low-power, non-ionizing radiation damages biological tissue. In contrast, ionizing radiation, such as the radiation from an X-ray, does have the potential to directly damage tissue.

Years of research have disclosed that, under certain circumstances, digital portable phones may interfere with cardiac pacemakers and hearing aids. However, despite years of research by several independent organizations, it is still not totally clear whether or not there is any risk of brain cancer from the extended use of cellular phones.

Nevertheless, British scientists have demanded that mobile phones carry a health warning. Roger Coghill, who runs an independent research laboratory in Wales, has gone so far as to say that "anyone who uses a mobile telephone for more than twenty minutes at a time needs to have their brain tested."

Virtually all lawsuits against cellular phone manufacturers and carriers have failed, usually due to a lack of clear scientific evidence that using a cell phone is dangerous to anyone's health. Still, specific concerns remain as research in this area continues.

Possible brain damage isn't the only issue. An industry-sponsored research program has determined that some digital phones caused interference in more than half of the 975 pacemaker patients tested. One of the conclusions of this study is that anyone with a pacemaker should use an analog rather than a digital phone.

More than 90 percent of phones used in the United States are analog phones, although that number is skewed by the growing number of new digital cellular and PCS phones. Most important, there is no actual health risk in using a wireless phone while wearing a hearing aid.

Digital phones do, to some degree, have the potential to interfere with hearing aids. But there are also some hearing aids that are immune to digital wireless devices. Wearers of nonhardened hearing aids can use analog wireless phones with no interference. They are also not likely to experience interference from someone else's digital phone, for instance, while walking near a digital phone that is in use.

The FCC long ago adopted safety standards for cellular phone transmissions and other devices that operate in a similar frequency range, and cellular phones fall well within the safety range. In 1997, FCC officials emphasized at a national summit convened on radio frequency (RF) exposure that its rules do not limit the emissions of any transmitter, but instead put limits on human exposure to RF fields in excess of the maximum permissible exposure (MPE) limits. The FCC also suggested several measures to protect workers and the public from being exposed unnecessarily to RF emissions, such as:

- Restricting access to a transmitter site

- Elevating antennas above head level

- Implementing a comprehensive site safety program

- Requiring all workers to use personal monitoring devices while at the site

• Mandating the use of protective RF-shielding clothing when working in high RF fields

The Center for the Study of Wireless Electromagnetic Compatibility at the University of Oklahoma, which is chartered to work with the wireless and medical device industries as well as government agencies to resolve interindustry electromagnetic compatibility issues, has also developed a research protocol designed to quantify the source of any problem and recommend solutions.

The European Community (EC) has already recommended new hearing aid standards, making the devices more resistant to interference from all extraneous sources. And a recent study reported in the Australian Medical Journal suggested a link between the increased use of cellular phones and an increased number of brain tumors.

RF energy emissions from antennas have been another concern, but field tests by the FCC in 1992 indicate that measurements made around typical cellular base stations have shown that ground-level power densities are well below limits recommended by currently accepted RF and microwave safety standards.

Typically, antennas are mounted on a tower that is 100 to 200 feet high. The antennas are generally designed to radiate very high power directly downward, and exposure to electromagnetic energy decreases rapidly with distance. As a result, exposure at ground level tends to be small.

More recently, the U.S. Food and Drug Administration (FDA) recommended that more research be conducted into the possible health risks from mobile telephones. But the agency did not indicate that such a program is important enough to receive federal funding. In fact, the FDA notes that "there is no new information indicating that use of cellular phones is a human health risk."

Trade reports on the FDA statement suggest that there is disagreement within the agency about how to handle this issue, one of the more prominent being that the FDA wait to see how industry-funded research plays out.

Meanwhile, the Geneva-based World Health Organization has developed a plan to track mobile phone uses in eight countries over the next decade to determine whether portable wireless communications devices pose a health risk. The eight countries are Australia, Canada, Denmark, Italy, Israel, France, the United Kingdom, and Sweden.

Wireless Fraud

Much has been written about stealing cellular phone users' phone numbers to make free phone calls and other forms of wireless fraud. With technology advances, education, and law enforcement, the cloning of cellular phones is not as serious a problem as it used to be, but it is worth monitoring.

Actually, there are a number of techniques for making fraudulent calls, but essentially this occurs when criminals take apart a wireless phone and reprogram it with a counterfeit account code, which tricks a wireless system into sending the bill elsewhere.

The customer should not get stuck with the bill. Virtually all wireless service providers have a policy that removes fraudulent charges from the accounts of customers. However, wireless fraud is not a victimless crime. It adds to the cost of doing business (the wireless industry claims that it lost $650 million in 1995 due to fraud, or about 3.8 percent of the industry's total revenue), and legitimate customers are inconvenienced since they must change their number when their phones are reprogrammed.

While most of the problem is in the bigger cities—mainly New York, Miami, and Los Angeles—wireless fraud can take place anywhere. Specifically, it is a violation of Title 18, Section 1029 to knowingly and with intent to defraud, use, produce, or traffic in one or more counterfeit wireless phones. On October 24, 1994, President Clinton signed H.R. 4922 (Communications Assistance for Law Enforcement Act) into law.

Amendments to Section 1029 now include the fraudulent alteration of telecommunications equipment. Punishment includes fines of up to $50,000 and fifteen years' imprisonment.

Additionally, the rules and regulations of the FCC prohibit tampering with and/or altering the electronic serial number (ESN) inside a wireless telephone. Every wireless phone must have a unique ESN and no two phones may have or emit the same ESN, according to FCC rules.

Cellular equipment manufacturers are now developing an authentication methodology to reduce fraud. This would replace the use of the ESN with an encrypted code that could not be obtained by off-the-air monitoring.

Meanwhile, until authentication can be put in widespread use, some cellular carriers have reduced fraud by requiring subscribers to use their personal identification numbers (PINs) before allowing access to the cellular system. Others are implementing various sophisticated fraud-detection technologies that help them to manage the fraud problem.

The U.S. Senate has also enacted legislation (the Wireless Telephone Protection Act) that strengthens criminal penalties against wireless phone cloners.

Frequently Asked Questions About Cellular Services

- *Why can't you use your cellular phone on an airplane?* It's an FCC rule, which states that cellular phones installed in or carried aboard airplanes, balloons, or any other type of aircraft must not be operated while the aircraft is airborne. All cellular phones on board must be turned off the moment the aircraft leaves the ground.

Curiously, while the rules governing the use of cellular phones while an aircraft is airborne are covered by FCC regulations (and are subject to a fine), the use of cellular telephones while the aircraft is on the ground is subject to Federal Aviation Administration (FAA) regulations.

- *What is prepaid cellular calling service?* Several phone manufacturers and service providers now offer prepaid calling service, allowing subscribers to prepay for airtime and value-added services. By prepaying for your mobile phone services, you can build up airtime credits that are automatically deducted in real time every time the phone is used, according to the call duration. At the start of each call, you're told how much credit remains. Value-added services such as fax and voice mail can also be prepaid as part of the package. You can continue to use up prepaid airtime even when roaming into other networks.

The main attraction of prepaid service to users is that it makes mobile phone ownership simpler, with no invoicing for users to attend to, and no contract to sign. For operators, it creates a completely new marketing opportunity, improves cash flow, and reduces exposure to fraud.

Prepaid services are also suited for low-usage users—for example, families who want a mobile service just for emergencies, or short-term subscribers, such as business travelers or tourists.

- *What is the Wireless Communications Service?* In late 1996, the FCC announced its intention to establish a new Wireless Communications Service (WCS) in the 2.3 GHz band. WCS is defined by the FCC as radio communications that may provide fixed, mobile, radio location, or satellite communication services to individuals and businesses within their assigned spectrum block and geographical area.

The FCC believes that WCS can provide more advanced wireless phone services that would be able to pinpoint a subscriber in any location, and it expects the creation of the WCS to lead to the development of an entire family of new communication devices using very small, lightweight, multifunction portable phones and advanced devices with two-way data capabilities.

WCS systems will be able to communicate with other telephone networks as well as with personal digital assistants, allowing subscribers to send and receive data and/or video messages wirelessly.

The FCC's licensing plan provides for several new full-service providers of wireless service in each market. Consumers and business users will be able to choose from multiple providers and are expected to be able to receive lower prices and better service as a result.

In July 1997, the FCC granted 126 licenses for WCS.

• *What are smart cards?* Smart cards, which store information electronically, have been in use for nearly two decades. The first application was a debit card for French banks and required insertion of the card into a terminal to transmit information. Today's smart cards use increasingly complex microprocessors with computing and memory storage. They can exchange information through contact with a reader or, with the newer contactless and combination smart cards, add a miniaturized radio modem for sending and receiving data via a radio frequency transmission.

Millions of smart cards have been introduced into the marketplace in recent years with predictions that they will replace nearly all forms of tender, including bills, credit cards, and checks. More recently, smart cards have found their way into wireless telecommunications, adding new and special features to cellular telephony.

Atlanta, Georgia-based BellSouth Mobility DCS, for example, offers international roaming with nine European GSM operators. DCS customers can use their subscriber identity module (SIM) cards (essentially the European form of smart cards) to roam overseas in Belgium, England, France, Ireland, the Netherlands, Spain, Sweden, and Switzerland. Service will be expanded to include Asia, Australia, the Middle East, and more countries in Europe. Because GSM operators in Europe use a different radio frequency than in North America (900 MHz and 1,900 MHz, respectively), DCS customers will be able to roam only by using their BellSouth Mobility DCS SIM card in a phone that operates at 900 MHz.

The SIM card contains details about a customer's service, including billing information and phone number. Callers can simply dial a customer's regular DCS number to be connected overseas. All calls made or received while roaming overseas are charged to a customer's regular DCS bill.

Phones that operate at 900 MHz can be rented in advance of traveling to Europe. By the end of 1998, dual-band phones that operate on both U.S. and European frequencies will become available, making travel outside North America easier.

CHAPTER 2

Fixed Wireless—
Hard Connections

I f you can pin them down, most business customers will tell you that what they want most from their telecommunications service is choice. In fact, most big businesses, and many small- or medium-sized business organizations, have a broad choice of telecom services today, and their list of options is growing all the time. They also want better service. And, of course, they want to pay as little for their telecom services as possible.

To a growing number of businesses, wireless communications systems provide a logical and cost-effective alternative to the traditional landline network. One way to do that is to provide basic, fixed telephone services. Wireless systems can often be installed in less time and at a lower cost, enabling service providers to provide universal telecom services where they didn't exist before.

Wireless Local Loop

The wireless communications industry is looking at every opportunity to compete with local exchange carriers (LECs), and many of them see the wireless local loop as the answer. Very simply, the local loop is the connection between the phone on your desk and the central telephone switching office. A wireless local loop (WLL) can be an attractive alternative to the installation of copper and fiber-optic lines. This is turning out to be especially true in regions of the world that do not have adequate telephone service.

In a typical wireless local-area network (WLAN) configuration, a transmitter/receiver (transceiver), called an access point, connects to the wired network from a fixed location using standard Ethernet cable. At a minimum, the access point receives, buffers, and transmits data between the WLAN and the wired network infrastructure. A single access point can support a small group of users and can function within a range of less than one hundred to several hundred feet. The access point (or the antenna attached to the access point) is usually mounted high, but it can be mounted just about anywhere that is practical as long as the desired radio coverage is obtained.

End users access the WLAN through wireless LAN adapters, which are implemented as PC cards in notebook computers, ISA (Industry Standard Architecture) or PCI (Peripheral Component Interconnect) cards in desktop computers, or integrated within handheld computers. WLAN adapters provide an interface between the client network operating system and the airwaves via an antenna.

In the United States and other highly developed countries, where landline service providers already offer quality voice and high data rate services at relatively low prices, WLL has been slow to grow. That's changing, however, as several cellular and PCS carriers have begun field testing a variety of fixed wireless services, some of which embody the features of both mobile and fixed (usually cordless) phones, with extensions of the wireless link out into campuses or even large neighborhoods. At the same time, several major telecommunications equipment manufacturers have begun to offer WLL premises systems in anticipation of integrating mobile and fixed services into a single service package.

Justifications for WLL

Fixed wireless networks are also becoming easier to justify with the construction of new buildings and office campuses in outlying areas. At the same time, wireline carriers are looking at using WLL to provide second lines to customers at a time when the demand for this additional service is skyrocketing, partly because of the growth of the Internet, but also owing to the demands for full-service small offices/home offices.

Figure 2-1 shows a LAN product from Lucent Technologies that is available for laptop, portable, and handheld computers.

WLANs can augment as well as replace wired LAN networks, often providing the final few feet of connectivity between a backbone network and the mobile users. The variety of applications is quite broad:

- Doctors and nurses in hospitals are more productive because handheld or notebook computers with WLAN capability deliver patient information instantly and without regard to the users' location.

- Network managers, using networked computers in older buildings, find that WLANs are a cost-effective alternative to running cable through asbestos-lined air ducts.

- Retail stores like the flexibility of using wireless net-

Figure 2-1. Lucent Technologies' WaveLAN IEEE 802.11-compliant PC card.

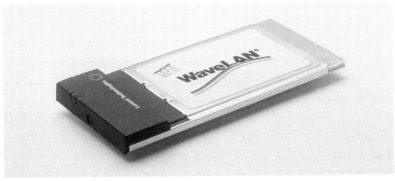

works to expand the number of point-of-sale (POS) terminals they can install on the sales floor, particularly during busy shopping seasons.

- Warehouse workers are using WLANs to exchange information in real time with central databases, increasing their productivity.

- Senior executives in conference rooms are making quicker, more informed decisions because they have access to real-time data at their fingertips.

The International Picture

Installing fixed wireless services in most underdeveloped and developing countries is a no-brainer. The numbers are sometimes hard to comprehend. At the beginning of 1997, 62 percent of all main telephone lines were installed in just twenty-three countries. Eighty-four percent of mobile cellular subscribers, 91 percent of all fax machines, and 97 percent of all Internet host computers are found in developed countries. There are 5.6 billion people in the world, and only about 600 million installed phones.

Several of the underdeveloped economies are already spending heavily on WLL installations, with the expectation that more than 100 million WLL lines will be installed worldwide by the year 2000. In fact, some telecommunications industry analysts believe that WLL could account for more than half of all new lines installed around the world by 2003.

Several different technologies, including cellular, PCS, local multipoint distribution service (LMDS—essentially, a wireless cable service that operates at a very high frequency), digital electronic messaging service, digital broadcast service (DBS), and even the current local exchange carriers, are capable of providing WLL services. Any of these systems is capable of replacing the wired phone with a fixed or hybrid fixed/mobile communications system to provide basic local phone service, particularly in underdeveloped economies.

With the right system, WLL can also incorporate video-on-demand, interactive video, and data services.

In underdeveloped and developing countries, putting new copper in the ground for new installations is too expensive. Also, it takes too long and the copper often gets dug up and sold on the black market.

At least twenty-five of these so-called developed countries have already implemented fixed wireless access, or WLL telephony, systems. Major WLL deployments are underway in China, India, Indonesia, Sri Lanka, Colombia, Bolivia, Mexico, Brazil, the Philippines, South Africa, the Czech Republic, the Ukraine, and Russia. They are using traditional-looking telephones that sit on office desks or in homes, but they operate wirelessly and are linked through the local cellular network.

In most cases, these are not small systems. The United Kingdom had more than 30,000 WLL subscribers at the end of 1997. Bell Canada deployed a commercial WLL network under a temporary license at the beginning of 1998, and Australia is using WLL in rural areas with a second line for data applications in urban areas. WLL field trials are also underway in France. Kubrelecom of Krasnodar, Russian Federation is using digital CDMA technology for a WLL network in the Krasnodar region, where 250,000 names make up the waiting list for telephone service.

Between 2000 and 2005, the pace of WLL growth is expected to accelerate dramatically to about 200 million, with the increase in developing countries outpacing the growth of WLL installations in developed markets by a ratio of three-to-one. North America is projected to account for about 40 percent of the WLL market, followed by western Europe and Asia.

Closer to Home

Many service providers now offer wireless services in office, campus, and factory settings. In some cases these services are automatically linked to cellular networks, enabling users to communicate with their offices when they leave the premises.

Several companies are offering "supercordless" phones that can be used without incurring airtime charges at home or the office, but which convert to the public switched cellular network when the subscriber moves out of the normal range of a cordless telephone.

In the United States, AT&T Wireless Services, which has reviewed some sixty different WLL technologies in an effort to get back into the local service market, had been testing a fixed wireless system known initially in the telecom industry as Project Angel. Based to a large extent on field trials conducted by AT&T Wireless in Chicago, Angel was running behind schedule, apparently due to a few technical glitches and cost issues. For one thing, Angel is not very cost-efficient. (The cost of installation, switching, and the backhaul was running to about $1,000 per unit.)

AT&T's plan has been to use the licenses that it acquired through FCC auctions for the 10 MHz spectrum to provide its new fixed wireless service to customers. Angel would be offered as a replacement for the landline phones and could be based on a flat-rate billing, or could be used with personal base stations that would allow for zoned pricing. Significantly, the licenses cover more than 93 percent of the United States.

The AT&T system was designed—initially, at least—to provide users (initially consumers, as AT&T is going after the mass market) with two phone lines and the capability for high-speed Internet access at 128 kilobits per second. The system would let consumers use their wireless phones as extensions in their homes and offices at local service rates. They could use the same wireless phones for mobile services in areas that are served by AT&T's wireless network while paying mobile rates.

Bundled Services and Zone-Based Pricing

Most of the regional Bell companies, which already have access to spectrum, will likely take advantage of the already high penetration of cordless phones and expand that service into a more broad-based mobile service. US West, for one, is already providing wireless T1 links for wireless local loop services, and Southwestern Bell has conducted WLL field trials in St. Louis.

It is not unlikely that one or possibly two PCS licensees in each market will begin to focus on fixed wireless services when it becomes evident that five or six mobile communications per market are two or three too many. Another option for PCS operators, and one that many of them are investigating, is to create hybrid systems by adding fixed network services to their existing mobile systems.

The carriers have been working for some time to reduce the per-minute rate and offer bundled minutes of service in a simplified bill. The problem with this, according to many industry executives and analysts, is that bundling all wireline and wireless (including e-mail and Internet) services into one potentially very large bill might be too much to handle for many subscribers.

An alternative that is gaining some attention is something called "zone-based pricing," in which carriers offer a large bundle of relatively inexpensive minutes from a single cell in the network. When the subscriber moves outside the home area, calls would shift to a more standard mobile pricing scheme.

How Much Is It?

Over time, the increase in demand for 802.11-based WLAN products should boost competition and drive down WLAN costs. With WLAN nodes ranging from $500 to $3,000, depending on several variables, cost continues to be a key issue. The good news for WLANs is that, while their cost is coming down, copper prices are going up.

An influx of new players creating more competition and rapidly maturing economies of scale should continue to reduce the price of WLAN systems.

The question often asked is, when will telco service subscribers living in a typical metropolitan area in the United

States be able to choose between a wireline or wireless service? Many people believe they have come close to that now with their cellular or PCS phones in terms of available services and meeting their personal communications requirements. The only real issue now is how much they are willing to pay for a ubiquitous wireless service.

Expectations are high that the basic plain old telephone service (POTS) will soon be followed by some pretty advanced new systems (PANS). In fact, that process is well underway. A number of U.S.-based industry groups are working with the International Telecommunications Union (ITU) to develop medium speed—64 kilobits per second (Kbps)—and high-speed—128 Kbps—data services for the digital CDMA standard. Currently, 14.4 Kbps packet data service is available on WLL and mobile systems. At the same time, several programs are underway to offer multimedia applications over WLL.

The need is for higher speed data rates for applications requiring wireless connectivity at 10 Mbps (megabits per second) and higher. This will allow WLANs to match the data rate of most wired LANs. There is no current definition of the characteristics for the higher data rate signal. However, there is a clear upgrade path to the higher data rates while maintaining interoperability with 1 and 2 Mbps systems.

Several new products have become available that should attract business users, including a WLAN that combines voice traffic with a wireless Ethernet and corporate data networks and intranets, which could potentially slash long-distance telephone costs. This system, developed by Symbol Technologies. Inc., supports the 802.11 WLAN standard and has already been installed in at least 20,000 sites.

Key Players

Several companies, such as Lucent, Unisys, and Adicom Wireless, are offering CDMA-based WLL. Another company, WinStar, is deploying its services in the top twenty U.S. cities. WinStar has interconnection agreements with all of the regional

Bell companies, as well as GTE and several of the major independent local exchange carriers. In addition to its local carrier services, WinStar has long-distance authority in at least forty-seven markets, allowing it to offer a full package of telecom services in competition with other telecom service providers.

New York City-based Teleport Communications Group (TCG), acquired in 1997 by AT&T, operates in sixty-six major markets. TCG provides a WLL capability that enables it to economically connect customers to its fiber-optic networks, and to provide standalone broadband facilities where TCG doesn't have a fiber-optics network. However, TCG has no plans to go all-wireless and, in fact, continues to build out its fiber network.

In several areas of the world, Israel and Denmark, for example, mobile rates have already come close to—and in some cases are already equal to—wireline fixed rates on a per-minute basis. As a result, some people have actually given up their wireline service and are using only their mobile service.

In-Building Wireless Offices

If you have wireless communications at home—and more than 70 percent of American homes have cordless phones today—why don't you have it at work? In fact, in-building systems, usually an adjunct to public/private branch exchanges (WPBXs), Centrex, or key office phone systems, are already making significant inroads into many offices. These systems allow a standard cellular phone to act as a wireless extension to the user's desktop phone in private cellular networks. These systems usually offer simultaneous ringing of the desk and cellular phone.

Because of their demonstrated productivity gains and the mobility they offer employees, these systems are gaining in popularity in corporate America—so much so that telecom industry analysts are projecting that by the turn of the century, one out of five or six office phones will likely be wireless. By 2002, between 15 and 20 percent of the new PBX shipments are expected to be wireless.

In-building systems are typically deployed in unlicensed rather than licensed frequency bands because using a licensed band requires either purchasing a license, which could be expensive, or negotiating a usage agreement with a license holder. And you usually don't have to start over with a totally new phone system; PBX manufacturers are introducing wireless products that can work with the more than 65 million business phones in use in the United States today.

Justification for WPBX

By using a wireless PBX, a user can eliminate local phone company charges of a PBX located in another building. A company with several metropolitan branches can potentially save thousands of dollars a month in phone and data link charges, amortizing its investment in the wireless system in less than a year.

Most in-building wireless systems are designed for medium and large businesses and typically offer users all of the familiar PBX-like services such as call transfer, message indicator light, and conferencing, but on a wireless phone.

WPBX systems can use either cordless or cellular phones. Cordless PBX systems may use a proprietary technology, or they may be based on a wireless industry standard or protocol. These phones would operate in and around an office or manufacturing complex, but would not provide outside service. In cellular PBX systems, the same phone can be used with a WPBX within a company's campus environment as well as for general cellular service away from the facility. Because local cellular service providers are licensed to use the area's cellular frequencies, at least one of the local cellular carriers would probably have to coordinate the installation of a cellular-based WPBX. Cellular PBX phones operate just like cordless PBX phones within the boundaries of a business office, but there are no airtime charges with the WPBX service.

A typical system uses low-powered, inexpensive base stations connected to a central controller, which is linked to the company's PBX. Each base station usually provides coverage of a single floor in an office building or manufacturing facility.

There are several vendors offering in-building systems today, but most of the systems are integrated solutions offered by the major PBX vendors. However, with 9.7 million business users accounting for 34 percent of all cellular subscribers and 63 percent of all cellular revenue, most wireless carriers are expected to aggressively promote their "wireless office" products and services.

In-building wireless communications has several attractions. For one thing, they provide instant accessibility. People can find each other faster, making employees more productive and customers happier. AT&T Wireless Services believes that, over a year's time, the average employee in today's workforce spends 155 hours listening to voicemail, 202 hours attempting unsuccessful and delayed responses to pages and playing "phone tag," and 677 hours away from their desk phones without any immediate means of accessibility. Considering that the average work year totals 2,000 hours, more than half of that time is lost.

The benefits of an in-building wireless office service are:

- Cost reduction

- Increased productivity

- Improved and simpler communications

- Recovered time previously wasted in "voicemail jail"

- No juggling multiple numbers or handsets

These systems are also much more efficient than in-building paging systems, walkie-talkies, or pagers. With most systems, coworkers can reach each other using just four- or five-digit extension dialing on their wireless phone, whether they are dialing within the building or across the country. And because wireless in-building calls are routed through the company's PBX, there are no airtime charges—calls within the campus environment that use the on-premise microcell are usually charged a flat monthly rate. Also, new employees can be added to the system immediately.

Kmart Corp., for example, uses an in-building wireless telephone system in its Super Kmart Centers across the United States, enabling its managers and security staff, equipped with lightweight handsets, to much more easily and rapidly respond to calls for price checks and in-store and call-in customers.

Another example is a system installed by Georgia Hospital Health Services, a division of the Georgia Hospital Association, which connects staff members to the hospital's existing business communications network. The heaviest users are nurses trying to reach doctors and respond to questions about patients.

The Jacob Javits Convention Center in New York City has also installed a wireless PBX system so that conventioneers can rent portable handsets and make calls from anywhere inside the 1.5 million square foot facility.

Key Issues to Consider

Of course, capacity and cost are important issues. Typically, these systems range in capacity from about 80 to 1,500 phones and cost up to about $1,500 per user.

Compared to other wireless original equipment manufacturers (OEMs), there are actually few WPBX vendors. Key suppliers include:

- AG Communication Systems (Phoenix, Arizona)

- Alcatel (Richardson, Texas)

- Astronet (Duluth, Georgia)

- CTP Systems (Cupertino, California)

- Ericsson (Research Triangle Park, North Carolina)

- Lucent Technologies (Murray Hill, New Jersey)

- Nortel (Dallas, Texas)

- Panasonic (Secaucus, New Jersey)

- SpectraLink (Boulder, Colorado)

- Tadiran Telecommunications (Clearwater, Florida)

- Uniden (Fort Worth, Texas)

Others, such as AT&T Wireless Services, Hughes Network Systems (Germantown, Maryland), and Ericsson, are responding to this marketplace with new systems and services designed to enable mobile workforces to be reached anywhere, anytime, using cellular or PCS portables. These products are invariably compatible with major office PBXs and incorporate low power picocells throughout the building coverage area. Like basic cellular service, they feature seamless handoff of voice and data services between the internal network and external cellular services.

Increasingly, the carriers are bundling their airtime packages (usually flat-rate pricing for the office system) with the regular rate plans for mobile services.

Future Outlook

A big jump in capability and technology for WPBX systems and services could come in the next three to four years with the adoption of so-called Third-Generation (3G) concepts and technical standards by the International Telecommunications Union (ITU). One of the ITU's most important projects currently—and there is much more information on third-generation activities in Chapter 6 of this book—is International Mobile Telecommunications-2000 (IMT-2000), which has set as one of its goals the development of a standard for indoor, wide-area, and satellite operations that would allow for global roaming. New and evolving network designs are also anticipating the development of third-generation technologies and features.

Wireless Local-Area Networks

A wireless local-area network (WLAN) is another form of fixed wireless communications.

A WLAN provides a communications network that can be reconfigured without calling in technical specialists to rewire and reconfigure desktop computers, printers, fax machines, copiers, and videoconferencing equipment.

In a basic WLAN setup, a company's wired network would have wireless access points (an access point connects the WLAN to the wired network and is similar in function to a wired server) strategically located in areas where users would want to connect to the network wirelessly. These access points generally have a 10-BaseT Ethernet connection to the wired network, giving wireless users access to all network resources, including Internet access.

At the user level, a credit card-sized PCMCIA card or other interface module would enable laptop, desktop, or handheld computers to communicate with the access points, which relay information to and from the network.

When they were first introduced, WLANs were pitched as an alternative to wired, usually in-building or campus-wide networks. Offices are often reorganized, requiring phones and computers to be moved with the people. Thousands of dollars are spent in moving costs—often in old buildings with poor ducting that make rewiring difficult. Worse, the building may have been built with asbestos.

Companies don't have to scrap their wired systems and switch to WLANs. WLANs can be an add-on to wired networks, not a replacement. Fortune 1000 companies, for example, have an increasingly large number of mobile employees. It is estimated that more than 15 percent of current employees of large businesses spend most of their workday in or around their office buildings, but away from their offices and desktop PCs. It makes more sense to put up wireless access points for their mobile workers.

WLANs enable employees to sit down wherever they are in a building or the corporate campus, power up their laptop PC, and be connected to the company's network. The increased efficiency in employee productivity is significant.

Information technology (IT) executives are also discovering that separate buildings within a corporate campus that operate on a single network can often be connected much more cost-effectively by using a wireless bridge/router.

At the time the industry began pitching WLANs, some manufacturers, warehouses, and retailers were already using these wireless networks to link handheld terminals, enabling employees to collect and access data through a central computer from a remote site. WLANs are also being used in factory automation, inventory management, security, telemetry, and data acquisition systems, as well as in hospitals, stock exchanges, and real-time information access on corporate intranets. In addition, small offices, especially professional offices such as accounting and law firms, can benefit from networking their computers, and now can do so without investing in the time and complication of wiring their offices. A WLAN is an easy solution for their needs.

A warehouse installation would typically include a grid of fixed access points providing bridges to a wired Ethernet LAN. Mobile WLAN devices will register or "associate" with the nearest access point, transmitting their registration "seamlessly" to neighboring access points as the user moves around. Rental car companies are using WLAN systems to speed customers through the vehicle check-in process with wireless terminals (including printers) hanging from the belts of their service personnel. Restaurants are using similar terminals to report food orders to the kitchen from tables, and several have installed wireless terminals that allow customers to swipe their credit cards at their tables.

One of the most time-critical requirements for information exists on the trading floors of brokerage firms. Dealers on the trading floor are equipped with wireless personal digital assistants (PDAs) from which they can call up the latest financial information at the touch of a button. This family of applications relies on the fast response time and density of installation that modern WLANs provide. A trader wearing a wireless headset is shown in Figure 2-2.

Figure 2-2. An American Stock Exchange floor trader equipped with a wireless headset.

As more so-called vertical markets come into focus, markets have begun to develop for "horizontal" WLAN applications, which would essentially eliminate lap-link cables and plug-in floppy-disk drives. Exchanging data and presentation slides, for example, and even installing new applications, is possible by simply being in the proximity of an access point or a colleague's computer, or being within sight of a conference platform.

Few challenged the concept of a wireless LAN when it was first introduced, but the cost of installing a WLAN has been an issue. As the WLAN market grows, however, economies of scale will drive prices down and spur further investment in the technology. A bigger problem has been the lack of technical standards to govern WLANs. Until about mid-1997, potential customers could not buy compatible products from different equipment suppliers. They also worried about "betting on the wrong horse."

The IEEE 802.11, named after the Institute of Electrical and Electronics Engineers subcommittee that developed the standard, was formed in 1991 to standardize the emerging WLAN technology and to provide interoperability between systems from different vendors. Before the IEEE 802.11 definition, there were several proprietary solutions that did not provide interoperability.

Currently, the 802.11 standard defines WLAN communications in the 2.4 GHz frequency band at a maximum data rate of 2 megabits per second (Mbps). However, the IEEE Standards Activity Board is expected to update the standard to cover higher data rates and frequencies to make them more competitive with wired data transmission options. The target is to define WLAN data communications at 10 Mbps.

In fact, one of the key reasons that WLAN has been slow to win acceptance among business users was the lack of a technical standard. Now that one is in place, potential WLAN users are likely to be more comfortable with the technology, and they will have more options in terms of mixing and matching WLAN hardware and software from a variety of vendors. The 802.11 standard also helps pave the way for mass-market components that should help drive down WLAN prices.

Simply, 802.11 defines the physical (PHY) and medium-access-layer (MAC) protocols for WLANs. MAC works seamlessly with the Ethernet local-area network standard. The PHY specification encompasses three transmission options. One is infrared, of which there are few suppliers. The other two are RF-based options and include direct-sequence, spread spectrum and frequency-hopping, spread spectrum. The options are designed to cover a range of price/performance requirements.

Among the more popular WLAN products available today are IEEE 802.11-compliant PC Card radio modules for original equipment manufacturers of WLANs. The OEM PC Cards are designed for easy integration by system designers and developers and can even be installed in standard notebook and mobile computers.

High-Data-Rate Wireless Links

A recent development in WLANs is a wireless 11 Mbps technology that operates in the unlicensed 2.4 GHz band. The 11

Mbps WLAN promises a performance level comparable to an industry-standard 10 Mbps Ethernet system, with expectations that WLAN technology will soon be adopted by end-users as either a mainstream enhancement to wired networks or as a standalone networking solution, depending on the application.

The newer technology addresses the concerns of high-data-rate wireless links for practical business use. One is speed—that wireless networking is not as fast as current wired networks. The other addresses interference—that the newer WLANs will seriously disrupt communications or seriously impede performance. Tests have indicated, however, that high-speed data rates can be maintained even in an environment with microwave interference.

WLAN developments continue to progress rapidly and include a series of industry proposals for a new standard for multivendor interoperability that goes beyond the scope of 802.11. Called the interaccess point protocol (IAPP), it is aimed at establishing protocols that facilitate roaming across access points from multiple vendors. The 802.11 standard defines roaming, but only across access points from the same vendor.

Early in 1997, the FCC approved a plan to allocate 300 MHz of spectrum for unlicensed wireless data communications. This allocation of additional RF spectrum for unlicensed, high-speed data applications is aimed at providing new opportunities for next-generation communications technologies. By allocating additional frequencies for wireless communications, more WLAN products and services are expected to be developed, increasing communications options for businesses and other users.

Significant Applications

WLAN technology also lends itself to more commercial applications. The American Stock Exchange (AMEX) has launched a pilot program that will enable the more than 1,000 floor traders, brokers, and exchange staff to roam the 43,000 square foot AMEX trading floor while they trade options and equities using pen-based handheld computers. The heart of this system is a wireless LAN. The New York Stock Exchange has been testing a similar system. One bonus to the system is that it enables brokerage houses to monitor their employees on the floor—not only their transactions but also their work habits.

Halliburton Brown & Root (HBR), a global engineering and construction firm with clients in more than sixty countries, has been taking on shorter term, but more numerous projects. As a result, staff members are constantly on the move. Project teams form and re-form swiftly. The company knew that it needed a communications infrastructure that could accommodate its dynamic and flexible work style. After evaluating WLAN and cordless phone systems, HBR equipped a hundred notebook PCs with WLAN/PCMCIA cards and installed ten WLAN access points in its new headquarters building near London. Together these devices extended the company's Ethernet network capabilities to mobile users anywhere in the building's third and fourth floors. Each access point provides a 2 Mbps cell that connects mobile devices to servers, peripherals, and other PCs throughout the organization. The system also provides a high level of security Initially, the system was used by senior management. Increasingly, project teams have been using the WLAN network to send and receive e-mail messages, access company databases, and transfer files. One of the company's options under consideration is to set up a temporary WLAN at client sites or other temporary locations.

Another example of WLAN use is Illinois Power, which is developing a WLAN for an automatic meter-reading system. When in full operation, the system will provide coverage over the company's 15,000 square-mile service territory, connecting more than 600,000 utility customers. Eventually, the system will monitor more than one million residential, commercial, and industrial customers.

One of the more interesting WLAN applications is an in-grocery store advertising system called Klever-Kart. A very small WLAN terminal is mounted on food store shopping carts. As customers enter the store, the terminal on the cart is activated and promotional ads, coupons, and a store directory are transmitted over the RF network to the WLAN devices mounted on the cart. The system isn't cheap to install. How-ever, market data indicates that 80 to 90 percent of a person's buying decisions are made in the store. By using the WLAN system, food stores can direct the selection process and make better use of their advertising dollars.

Currently, Klever-Kart transmits only in one direction. In the future, two-way systems will be available and customers will be encouraged to swipe a frequent-shopper card through the cart WLAN terminal. Information on what the customer purchases regularly will be stored in a computer and store "specials" will be transmitted to the cart terminal display whenever the customer shops in a Klever-Kart-equipped store.

Another WLAN data device has been developed around a wireless thermostat. Many offices regularly move their cubicles around and re-size them to accommodate changes in staffing. The wireless thermostat can be attached with Velcro to a wall anywhere in a building and pulled off and moved to virtually any location as often as necessary. The unit communicates with other wall-mounted devices to maintain a balanced envi-ronment throughout an office or manufacturing facility.

Future WLANs will offer even more capability, including multimedia features such as wireless video transmission for mobile videoconferencing. However, as with most purchase decisions, the bottom line is the bottom line. In WLL, the pur-chase decision most often will be based on price and voice quality. Both of these issues are critical, as they should be, and will be based not only on the ability of WLL to provide a new and useful service, but how it stacks up in cost and quality with wireline services.

CHAPTER 3

Mobile Communications Satellites—Roaming the World

W here were you on May 19, 1998? If you own a pager, you may remember that day when a communications satellite spun out of control and interrupted service to millions of pager users in the United States. Literally millions of pager subscribers were affected by the glitch in the Galaxy IV satellite. Doctors and fleet service personnel had to resort to regular phone calls to stay in touch with their offices for at least a day. Parents couldn't contact their children. Police bomb squads could not be reached immediately in case of an emergency.

All of which points to the ubiquity of this service and how much so many of us have come to rely on this technology.

Now we have the prospect of huge networks of mobile communications satellites as the only wireless service available that offers true global roaming—that is, the ability to talk to anyone at any time, anywhere in the world. The service is unique, but who really needs it?

The business scenario might work like this: You're in Hong Kong, on your way to make a presentation that could result in a multimillion-dollar contract for your company and a sizable commission for you. Your staff back home in Atlanta has just come up with some new figures that could significantly improve your chances of winning the business, but they have no idea where you are at the moment. In fact, they don't even know where you're meeting the client. But they know that you have a portable satellite phone with you, and they have your number. All they have to do is call you—as if you were down the hall in your office—to update the information for your presentation.

Mobile Satcom Trends

More communications satellites are expected to be launched in the next ten years than all of those placed in orbit in the past thirty years. More than 145 communications satellites are already operating high above the equator in geostationary orbit, with at least another 155 GEOs expected to be launched by the year 2000. In addition, several hundred new low-earth orbit (LEO) mobile communications satellites will be launched over the next few years, along with several equally small satellites to cover a number of emerging regional communications satellite services.

As shown in Figure 3-1, approximately 1,335 communications satellites are scheduled for launch between 1997 and 2006. Mobile communications satellites will account for nearly half of these.

Most of the new mobile satcom (i.e., satellite communications) systems are targeting three users' groups:

1. International business travelers, particularly those who might work in areas where cellular coverage is poor or nonexistent

2. Heavy cellular phone users who frequently operate out of their normal coverage areas

3. Mobile telecom users in underdeveloped and develop-

Figure 3-1. Projected communications satellite launches, 1997–2006.

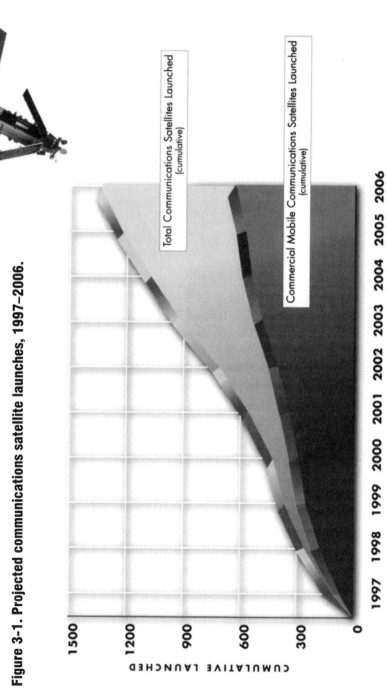

Total Communications Satellites Launched
(cumulative)

Commercial Mobile Communications Satellites Launched
(cumulative)

CUMULATIVE LAUNCHED

1500

1200

900

600

300

0

1997 1998 1999 2000 2001 2002 2003 2004 2005 2006

CUMULATIVE LAUNCHED BY YEAR

Source: The Teal Group, Fairfax, Virginia.

ing countries with little international telecommunications infrastructure

In developing economies of the world, these new satellite systems will support private portable and fixed phone services where they did not previously exist (e.g., in businesses and homes), as well as strategically placed telephone kiosks (i.e., Apay stations), providing even the remotest village in the world with a full range of global telecom services.

The opportunities for these new services are immense. Half of the people in the world have never made a phone call. More than two-thirds of the world's geography is without wireless services. That could change quickly as these new and emerging mobile communications satellite services begin to provide not only voice, but also data and video services to these largely unserved or underserved regions, including data acquisition and monitoring applications that cannot be served by terrestrial communications.

Early Adopters: GEO Users in the Transportation Industry

The first truly mobile geostationary orbit (GEO) system in operation is Reston, Virginia-based American Mobile Satellite Corp. (AMSC). In 1989, the Federal Communications Commission (FCC) granted AMSC a license to provide a full range of mobile satellite services to the U.S. land mobile, marine, aeronautical, and fixed-site markets. Rather than launch its own fleet of satellites, AMSC has leased capacity from INMARSAT, providing mobile data and position-location services to the transportation, maritime, rail, and remote monitoring industries.

AMSC, whose shareholders include Hughes Communications, Singapore Telecom, and AT&T Wireless Services, began to operate several of these services in late 1995, including a dual-mode (i.e., satellite/cellular) phone service and a mobile messaging service designed specifically for transportation companies.

AMSC got another jump on the market at the end of 1997, when it announced its acquisition of Motorola's ARDIS, the largest two-way wireless data communications network in the United States, covering more than 425 of the top cities in the United States, Puerto Rico, and the United States Virgin Islands. The deal will make AMSC one of the country's largest providers of mobile data services to the transportation and field services industries, as well as others that rely heavily on mobile communications. The AMSC network will have the only combined satellite/terrestrial data capability for least-cost routing of data messaging.

In January 1998, seeking to gain new market segments, AMSC paid Motorola $100 million to acquire ARDIS, the largest two-way data network in the United States. Covering some 425 major cities, ARDIS is used primarily by businesses such as trucking companies to send and receive data to drivers on the road.

Another GEO satellite is in development in Japan that is expected to be adopted by that country's cellular carriers to fill in the cellular coverage gaps in Japan. Once the system is in place, handheld mobile phone subscribers would be able to communicate with each other from anywhere in Japan.

Big LEOs

The newest trend in mobile satellite services involving networks is low-earth orbit (LEO) satellites. There are two types of LEOs:

1. *Big LEOs.* These satellites provide both voice and data services and operate above 1 GHz.

2. *Little LEOs.* These satellites provide data services, including mobile messaging and tracking/location/ monitoring, but no voice communications, and operate below 1 GHz.

The Big LEOs include:

- Iridium (headed by Motorola, but with a long list of investors), with handheld telephones (see Figure 3-2) capable of delivering voice, fax, and data services anywhere in the world and offering terrestrial cellular interconnections with automatic switching to the Iridium global satellite communications network if cellular is not locally available

Figure 3-2. The satellite-based Iridium handheld telephone.

- Globalstar (owned and operated by a Loral/Qualcomm joint venture)

- Teledesic (led by Microsoft's Bill Gates and Craig McCaw, who sold McCaw Cellular Communications to AT&T)

- ICO Global Communications (owned by a diversified list of communications companies)

- Ellipso (Mobile Communications Holdings)

- ECCO (Constellation Communications)

These services are in different stages of design and deployment, although Iridium is likely to be the first system in service.

Regulatory Trends and Intergovernmental Satellite Operations

Another Big LEO, the TRW-sponsored and -developed Odyssey, was merged into the ICO program when London-based ICO Global Communications acquired a major stake in the TRW program in January 1998. ICO was established in early 1995 as a private company to provide satellite-based personal mobile global communications services. Once it is on the air, which is scheduled for the year 2000, ICO says that its network will be capable of connecting a subscriber using portable phones to any telephone around the world.

However, the TRW/ICO deal has muddied the international regulatory waters by creating a U.S. partner for a European-based satellite phone system. Prior to the acquisition, ICO was the only foreign challenger to the other Big LEOs. Although a private company, ICO was developed as a spin-off of INMARSAT. Consequently, U.S.-based mobile satcom contenders claim that ICO has an unfair regulatory and financial advantage because it evolved out of INMARSAT, which has been operating globally since 1979 through an international treaty. ICO's position is that it is a totally separate business entity.

To better understand regulatory trends, let's first clarify the various organizations involved in the move to privatization:

- COMSAT. The U.S. Congress created the Communications Satellite Corp. as a private company in 1962. COMSAT provides communications capacity to U.S. carriers such as AT&T and others.

- INTELSAT. The International Telecommunications Satellite Organization is an international-formed organization that is involved in launching and operating commercial satellites. More than 100 nations participate in INTELSAT, which is partially owned by COMSAT.

- INMARSAT. The International Maritime Satellite Organization is another international organization that provides satellite communications to and from ships and offshore rigs. It is represented in the United States and is partially owned by COMSAT.

COMSAT, as INMARSAT's U.S. signatory with a 22 percent equity position in the international satellite organization, has called for the privatization of intergovernmental satellite organizations and the deregulation of its own operations.

COMSAT's initial argument is that INTELSAT and INMARSAT both face anomalies in their structure and the way they are governed. Both face intense competition from undersea fiber-optic cables and other satellite systems and believe they must privatize to compete. To do this, INTELSAT and INMARSAT have begun to restructure themselves toward privatization.

COMSAT is quick to point out that between early 1994 and early 1998, the satellite industry saw the creation of twenty-one public companies representing at least $14 billion in investments. However, a similarly impressive growth has taken place in undersea fiber-optic cables. Today, taking satellites and fiber-optic cables together, there are multiple alternatives for 98 percent of the circuits to and from the United States.

COMSAT claims that, unlike itself, its satellite competitors are not subject to any FCC economic regulation and that if it were deregulated, it could improve its efficiency and be quicker to market with new services and advanced technologies.

INTELSAT further complicated its international regulatory status by announcing plans to create another spin-off company targeting interactive multimedia services to business and residential customers. To be called INC, the new company will operate separately from INTELSAT with its own management team and board of directors. INC will work with only six satellites.

By mid-1998, plans to privatize INTELSAT were well under way. As part of the plan, COMSAT Corp. was deregulated by the FCC in the major markets of its largest business unit, COMSAT World Systems. This action deregulates services that account for about 90 percent of this unit's revenues. The action liberates the corporation from utility-style earnings regulation and finally allows COMSAT to compete on a more equal footing with other international telecommunications and satellite companies.

The FCC also eliminated the long-standing structural separation regulations previously applied to COMSAT World Systems, which provides service via INTELSAT satellites.

Proposals wending their way through the regulatory process in mid-1998 would make COMSAT a shareholders and board member in the new private INTELSAT. COMSAT's ownership in INMARSAT would be transferred to the new company, which would make a public offering within two years after privatization.

Pricing: (Satellite) Talk Is Not Cheap

Most telecom industry analysts believe that mobile satellite services will play a small but important role in bridging the connectivity gap during the global buildout of telecommunications networks. Whether the LEOs succeed or not will depend to a large extent on how the need for this new service is perceived, and how much users are willing to pay for the service.

Using a Big LEO will be more costly than using a cellular phone. The satellite phone itself could initially run to more than $2,500, while the initial airtime charge for using Iridium could be as high as $3 a minute. Globalstar says that it expects its service to begin at $0.47 a minute.

Much of the difference in pricing has to do with the way the systems work. Although Globalstar calls may originate in remote areas, the system should be able to switch into a ground-based network within 1,000 miles of where the call originated. Iridium, on the other hand, sends its signals from satellite to satellite and terminates as close to its destination as possible before connecting to a public-switched telephone ground station.

An analysis by CIBC Oppenheimer puts forth two scenarios for making a call from Australia to the United Kingdom:

• *Calling route 1.* Using Globalstar, the call is received by a satellite in the Australian region, then transmitted to the nearest gateway in Australia. Globalstar charges the service provider $0.47 per minute. From the gateway, the call is routed through the public-switched telephone network (PSTN) from Australia to the U.K. The long-distance carrier bills the customer (through the service provider) at its per-minute rate. Since a call from Australia may have to access several different countries, it will be more expensive than the call from India to the U.K. The rate is about $1 per minute. In effect, the cost to provide the service is $1.47, which the service provider can mark up for a profit.

Figure 3-3. The Iridium personal pager.

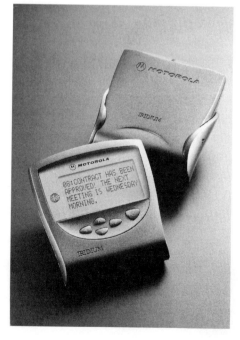

• *Calling route 2.* Using Iridium, the call from Australia to the U.K. is received by a satellite in the Australian region and transmitted from satellite to satellite to the gateway nearest the person receiving the call in the U.K. In this case, a gateway in Italy would be used. Iridium would charge the service provider $2.50 per minute. From the gateway, the call would be routed through the PSTN from Italy to the U.K. The long-distance carrier would bill the customer (through the service provider) at its per-minute rate. From Italy to the U.K., that rate would be about $0.12 per minute. In effect, the service provider's cost is $2.62, which it can mark up for a profit.

Iridium additionally offers global roaming capability in a small, belt-worn personal pager (see Figure 3-3). These pagers provide service in areas with incompatible terrestrial paging systems or no terrestrial paging coverage.

Survival for the Fattest: Competition for High Bandwidth

Of course, time and competition will drive down the cost of these services, probably at a fairly rapid rate. United States and European companies have already announced plans to place a total of nearly 250 satellites in orbit to support narrowband voice and data services.

Another factor to consider is that, while satellites take distance out of the cost equation, the ground-based carriers may at some point attempt to restructure their pricing for mobile satellite telecom traffic.

At least one study, called Next-Generation Media Technologies, by Menlo Park, California-based SRI Consulting, suggests that even if only a fraction of the planned satellite communications projects become operational over the next several years, it will be enough to spark a price war for all wireless voice services.

The SRI study also points out that global demand for voice and data networking means that the world of satellite communications may no longer revolve only around North America. Satellite arrays will connect all regions of the world.

Even the Geneva-based International Telecommunications Union (ITU), which allocates the scarce radio frequencies for communications satellites and other wireless services, projects tough competition for the Big LEOs. These new systems will require the assignment of new frequencies, and the ITU is on record as claiming that there simply is not enough available spectrum for all of the proposed new services. The fact that at least a dozen companies hope to provide satellite-based high-bandwidth Internet and video services beginning in the year 2000 will only aggravate the issue.

Obviously, it is too early to make clear judgments on the success of these high-bandwidth multimedia services, but it is highly unlikely that the larger cable television and telephone companies will sit still and let this premium-level business be taken away from them.

Who's Who: Leading Mobile Satcom Players

Teledesic

The largest of the mobile satellite networks is the $9 billion Teledesic program. Founded by cellular industry pioneer Craig McCaw and Microsoft's Bill Gates, with a more recent investment by Motorola and Prince Walid bin Talal of Saudi Arabia, this Internet in the sky system will require 288 LEO satellites (the original plan called for 840 satellites) to be operational by the year 2002.

How Teledesic Works

Rather than selling its services directly to end users, Teledesic will provide an open network that will enable local telephone companies and government agencies in host countries to extend their networks, both in terms of geographic scope and in the kinds of services they can offer. Ground-based gateways will enable service providers to offer seamless links to other wireline and wireless networks. Fixed and mobile user terminals will communicate directly with Teledesic's satellite-based network to other terminals and through gateway switches to other networks, such as the public switched telephone network and the Internet.

The network uses fast packet-switching technology, with a packet design similar to asynchronous transfer mode (ATM). All communication is treated identically within the network as streams of short fixed-length packets. Each packet contains a header that includes address and sequency information, an error-control section used to verify the integrity of the header, and a payload section that carries the digitally encoded voice or data. Conversations to and from the packet format take place in the terminals. The fast packet-switched network combines the advantages of a circuit-switched network with low-delay digital transmissions and a more efficient packet-switched network.

The Teledesic network will accommodate a wide variety of data rates.

Iridium

However, the first Big LEO to be operational will be Iridium, a network of sixty-six satellites developed and managed by Motorola. Iridium is expected to begin formal operations in September 1998. Iridium was originally conceived as a system of seventy-seven satellites—hence the name Iridium, for the element whose atom has seventy-seven orbiting electrons. Motorola engineers were able to reduce the number of satellites in the system to sixty-six and still cover the entire earth by configuring them in six polar orbital planes of eleven satellites each. Each of these satellites will orbit about 420 nautical miles above the earth.

Complicating both the competitive and regulatory outlook for the Big LEOs, a consortia that includes Iridium and other major mobile communications satellite services (including Globalstar LP and Mobile Communications Holding) is developing plans for a more ambitious second-generation network of satellites with dramatically enhanced performance. The new Iridium system would require ninety-six satellites and would be able to send voice and data up to eighty times as fast as the sixty-six-satellite network it began launching late in 1997. Iridium has told the FCC that it would like to have this second-generation system in operation by 2002.

How Iridium Works

When an Iridium telephone is activated, the nearest satellite—in conjunction with the Iridium network—automatically will identify the caller account and the location of the user. The subscriber will select among cellular or satellite transmission alternatives, depending on the compatibility and system availability, to dispatch a call.

If the subscriber's local cellular system is unavailable, the telephone will communicate directly with a satellite overhead. The call then will be transferred from satellite to satellite through the network to its destination, either another Iridium telephone or an Iridium ground station. Iridium system gateways interconnect the satellite network with land-based fixed or wireless infrastructures worldwide.

ICO Communications

Initially referred to as Project 21 (for the twenty-first century) and then INMARSAT-P (for INMARSAT-Portable), ICO Global Communications (ICO stands for intermediate circular orbit) is a twelve-satellite system that would provide all the services planned by other mobile satellite systems. ICO Global Communications has proposed rates that would make it competitive with basic cellular service, but would also offer paging, facsimile, and data services beginning in the year 2000.

ICO's basic terminal will be a handheld unit similar to a cellular phone. It will be dual-mode (cellular and satellite) and will operate with the European GSM digital cellular technical standard, as well as with U.S., Japanese, and other cellular standards. ICO also expects to introduce some special-purpose terminals to handle dedicated data, mobile, maritime, aeronautical, and supervisory control and data acquisition (SCADA) units that are used in remote sensing and data collection applications.

ICO has fifty-seven investors, comprising telecommunications and technology companies throughout the world. They include a number of the world's top telecommunications companies collectively serving around 25 percent of the global cellular services market. More than half of ICO's investors are from developing markets.

Celestri System

Motorola and Matra Marconi Space S.A. had also announced plans to jointly develop a broadband satcom network, called the Celestri System, to handle mostly wireless data. Under their agreement, Matra Marconi Space would design and manufacture the satellite platform for use in Celestri satellites that are being designed and built by Motorola. The plan was to develop these systems for seventy LEO satellites and on GEO satellites.

However, rather than compete with a similar service, Motorola dropped its plans to develop Celestri in mid-1998, opting instead to back Teledesic with an initial investment of at least $75 million. Most industry analysts supported the Motorola move, suggesting that Celestri might be at least one satellite network too many for the marketplace to absorb and support and that Motorola, with its well-publicized financial and management/organization problems in 1998, probably should not be involved in another satellite program as ambitious as Celestri.

Globalstar

Globalstar, led by Loral Space and Communications Ltd. and Qualcomm, Inc., with ten other investor companies, also has proposed a second network of sixty-four satellites to supplement its current system in high-traffic regions. Globalstar began launching the first of its forty-eight first-generation LEO satellites in December 1997.

How Globalstar Works

The Globalstar system of forty-eight satellites is fully integrated with existing fixed and cellular telephone networks. Globalstar's dual-mode (i.e., satellite and cellular) handsets will be able to switch from conventional cellular telephony to satellite service when necessary.

Globalstar will enable international travelers to make and receive calls or faxes directly through their satellite-based mobile handsets anyplace in the world where Globalstar service is authorized by local regulatory authorities. Globalstar's dual-mode handsets are expected to retail for about $750. A Globalstar handset is pictured in Figure 3-4.

Globalstar's plan is to sell access to a worldwide network of regional and local telecommunications service providers, including its strategic partners. AirTouch Communications,

Figure 3-4. A Globalstar satellite-based mobile handset at work.

Dacca/Hyundai, France Telecom/Alcatel, and Vodafone in the U.K., together with Elsag and Loral, have agreed to act as Globalstar service providers in more than ninety countries. Each of these service providers will have the exclusive right to offer Globalstar service in its operating areas and will market and distribute the service, obtain all necessary regulatory approvals, and own and operate the gateways needed to serve their respective markets.

Ellipso and ECCO

Another Big LEO is Mobile Communications Holding's Ellipso system of seventeen satellites. Ellipso is scheduled to be in commercial service in the year 2000. However, the company wants to raise the stakes with a new twenty-six-satellite system capable of handling up to three times the voice and data traffic of its original seventeen satellites.

ECCO, an eleven-satellite Big LEO network operated by Constellation Communications, Inc., is scheduled to be in operation in 2000.

Spaceway

Hughes has taken a somewhat different approach with the design of a nine-satellite GEO system called Spaceway, which is aimed at offering an interactive bandwidth-on-demand service for fixed telephony, high-speed data, and high-resolution video. Spaceway will operate at a very high frequency of 28 GHz with tightly focused spot beams to accommodate the use of relatively small, 26-inch antennas.

Hughes says that it will price the satellite dishes under $1,000, or only slightly more initially than 18-inch direct broadcast satellite (DBS) TV satellite dishes. In fact, if Spaceway operates as advertised, it will provide two-way video transmission that is competitively priced with international phone calls and less than domestic phone calls.

Hughes sees two markets for Spaceway. One is similar to the Big LEO systems, providing basic telephone services to the telecommunications-poor regions of the world. Hughes' second target is the global business marketplace, for which services would include:

- True, full-motion desktop videoconferencing

- Computer-aided design workgroup computing for manufacturers

- Technical and medical tele-imaging (with the capability to transmit an X-ray every eight seconds)

- Low-cost, high-speed access to the next generation of online multimedia databases

The Five Little LEO Systems

At this point, the FCC has divided up frequency allocations to five Little LEOs. One of these, Orbital Communications Corp. (ORBCOMM), a unit of Orbital Sciences, has also asked the FCC for permission to build and operate a global satellite network that would provide broadband fixed-site communications services. This new service, called OrbLink, would offer high-speed data-transmission services at a fraction of the cost of current terrestrial and satellite alternatives and would be based on only seven satellites in MEO (medium-earth orbit) equatorial orbit.

The other Little LEO candidates for launch include:

- E-SAT, Inc., which is 80 percent owned by EchoStar Communications Corp., and 20 percent by DBS Industries, Inc.

- Leo One USA Corp.

- Final Analysis Communications

- Volunteers in Technical Assistance, a nonprofit entity that provides technical assistance to developing nations

Even if all of these proposals win licenses from the FCC, they must also secure licenses from the governments of foreign countries where they hope to operate.

How ORBCOMM Works

In the ORBCOMM system, a message is sent from a remote subscriber communicator unit in the United States—either stationary or mobile—and is received at the satellite and replayed down to one of four U.S. gateway earth stations. The gateway then relays the message via satellite link or dedicated terrestrial line to the network control center (NCC). The NCC routes the message to the final addressee via e-mail, dedicated telephone line, or facsimile. Messages originated outside the United States are routed through gateway control centers in the same manner.

Planet 1 and COMSAT-C

Just as American Mobile Satellite Corp. is already well entrenched in the GEO market, COMSAT Mobile Communications is making some headway in the mobile market through Planet 1, a PCS-type service that uses a portable, computer laptop-size satellite phone to provide voice, fax, and data communications services. However, while COMSAT's Planet 1 can carry inbound calls from customers virtually anywhere in the world, it does not have regulatory authority to provide outbound service from some countries, including the United States.

COMSAT views Planet 1 as a stepping stone to a much larger position in the world mobile satcom market. Various industries working in remote regions of the world, such as petroleum and mineral interests, have found Planet 1 to be very useful. COMSAT is expected to extend Planet 1 service into a larger system with more features, but with smaller mobile or even portable terminals.

COMSAT also offers global coverage through COMSAT-C, a data-messaging service. Through an alliance with T-Mobil,

Germany's largest supplier of mobile communications services, COMSAT-C customers can now send and receive data messages worldwide.

SkyBridge

Other international players include Toshiba Corp. of Japan, which has agreed to invest $3.5 billion in the SkyBridge Limited Partnership satellite project proposed by Alcatel Alsthom of France and Loral Space and Communications Corp. of the United States. SkyBridge is based on a constellation of sixty-four LEO satellites with downstream speeds similar to terrestrial broadband technologies. SkyBridge is scheduled to begin operation in 2001 with services delivered through local telecom carriers.

Location-Tracking Systems: Where Am I?

At some level, the Little LEOs, with their position-location capabilities, will compete with a number of currently available satellite-based services that provide similar services for long-haul trucking and other commercial vehicle operators.

OmniTRACS

Perhaps the largest and most successful of these is Qualcomm, Inc.'s OmniTRACS, a two-way text messaging and position-location network that is being used by at least 60,000 trucks in the United States, as well as fleets of commercial vehicles in Europe (where it is called EurtelTRACS and is managed by Alcatel in a joint venture with Qualcomm) and Japan (where it is operated by Qualcomm KK in a partnership with C. Itoh & Co., Nippon Steel Corp., Clarion, and Maspro, a supplier of DBS receivers). OmniTRACS leases channels on GTE Spacenet satellites for OmniTRACS.

The OmniTRACS system allows truckers to keep in touch with their dispatchers through an alphanumeric display terminal with a full keyboard in their cabs. The system virtually eliminates time-consuming and costly phone calls to the home office, provides drivers with accurate directions to their destinations, monitors in real time (via sensors) the status of refrigerated trucks, and immediately informs drivers of cancellations of pickups (obviously saving time and money). It also virtually eliminates the chance that their trucks will be successfully stolen. In addition to making trucking companies and their drivers more productive, OmniTRACS also serves as a type of personal national paging system for drivers in that dispatchers can tell them at any time to phone home. Delta Airlines has also used OmniTRACS on vehicles used to transport jet engines, and the United States Navy and Coast Guard use the system to monitor certain types of cargo.

Global Positioning System: From Military to Mainstream Uses

In fact, it is now fairly easy to determine your precise location in the world—to within 100 meters or less through the use of the U.S. Department of Defense -owned and -operated Global Positioning System (GPS), a network of twenty-four orbiting satellites designed specifically for global navigation.

Who would use GPS? Obviously the military, which has been using its GPS since it went into operation in the 1950s. GPS received a great deal of attention in the Persian Gulf War when soldiers on the ground used handheld GPS terminals to pinpoint their locations (in longitude and latitude) in the featureless desert. It received so much attention, in fact, that worried parents of soldiers responded to news that military GPS receivers were in short supply by purchasing commercially available handheld GPS models for as much as $2,500 and rushing them off to their soldier-children serving in the Gulf.

But the GPS system is also available for commercial use. Hertz Rent-A-Car is having some commercial success renting

GPS-based in-vehicle navigational mapping systems that help drivers keep from getting lost in unfamiliar territory. Some of these systems use GPS in combination with a gyro-heading sensor and wheel-revolution counter to determine the vehicle's position to the closest street. The database that provides the mapping information is stored either on a CD-ROM or a wallet-size PC card that plugs into the GPS receiver.

GPS-based car navigation systems have been in use in Japan for at least ten years, and the systems are gaining in popularity worldwide with a growing number of recreational users (e.g., boaters and hikers), as well as for such commercial applications as precision mapping, surveying, and asset and fleet tracking. Salem, Oregon-based II Morrow, Inc. has developed a custom portable GPS system for United Parcel Service (UPS), which owns II Morrow.

The U.S. Department of Transportation's Intelligent Transportation Systems (ITS) Joint Program Office estimates that there are at least 250,000 vehicles in the United States, including 100,000 trucks, using GPS. These numbers will increase dramatically as automakers such as Oldsmobile, BMW, Honda, and Toyota have more success with the optional navigation systems in their new top-of-the-line vehicles.

Market researcher Forward Concepts expects the worldwide commercial GPS receiver market to grow from $1.8 billion in 1995 to nearly $10 billion in the year 2000. Much of that growth, most analysts agree, will come from integrating GPS functionality into other portable products, including cellular and PCS phones, as well as laptop and notebook computers and pagers.

Augmented GPS

Several vehicle-mounted GPS/cellular phone systems are currently available from a number of manufacturers. One problem with using GPS in laptops is that the card slot in these and other devices, as well as the 232-standard serial port, is often spoken for with a modem, mouse, or some other connected peripheral device. A less significant but sometimes annoying issue is that vehicle-mounted GPS/cellular systems require two antennas—one for the cellular phone, the other for GPS.

The Federal Aviation Administration (FAA) is augmenting the GPS system with additional satellites, ground stations, ground-based transmitters, and monitoring receivers (called pseudolites), and control software in order to improve the integrity of the system. Two key improvements in the system are the Wide Area Augmentation System (WAAS) and the Local Area Augmentation System (LAAS).

WAAS supplements the twenty-four-satellite GPS constellation with several geostationary satellites that provide additional GPS-like signals for position measurement while broadcasting correction information and warning messages. Monitoring receivers are also being developed for LAAS to augment the GPS system for aircraft landings in poor weather.

International Navigation Programs

The Clinton administration has made the GPS system more attractive for commercial and consumer applications by ordering the discontinuance of a process called selective availability (SA) sometime after the year 2000. SA allows the Pentagon to intentionally degrade the accuracy of so-called public sector signals provided by the GPS network. Nevertheless, a number of countries (and companies) have expressed their concern about being tied into a system that is owned and managed by the U.S. Defense Department. At some point, most of these nations are expected to participate in the United States National Satellite Test Bed Program, and they will work through the International Civil Aviation Organization's Global Navigation Satellite System Panel to develop a worldwide WAAS signal standard.

The European Geostationary Navigation Overlay Service, Japan's Geographical Survey Institute, and the Arab Institute of Navigation, among other, similar organizations, have developed or are developing their own GPS-type navigation systems or are working to create enhancements to the U.S. system.

The only major alternative at the moment is Russia's twenty-four-satellite Global Navigation Satellite System (GLONASS), which could play a larger role in international navigation programs if manufacturers begin designing portable satellite-based navigation systems capable of using the Russian system. However, GLONASS has never quite measured up to the U.S. Global Positioning System technically, and Russian authorities have been unwilling to clearly indicate their ongoing support for the system to the extent that commercial equipment manufacturers have been willing to design, develop, and manufacture GLONASS-based hardware on a significant scale.

CHAPTER 4

Wireless Data—More Than Just the Internet

Wireless communications equipment manufacturers and service providers continue to look wistfully at market projections for wireless data services and wonder when they will start to kick in. At first glance, the numbers are impressive.

Out of the 125 million people employed in the United States, an estimated 40 million are mobile workers, and that number is expected to surge by as much as 50 percent of the working population by the year 2001. Even if the workforce doesn't grow over the next three to four years, industry consultants expect the percentage of mobile workers to grow from 32 percent to 50 percent. That means that 62.5 million workers would be wireless voice and data customers by 2001.

Slow User Acceptance of Wireless Data

Of course, those are the projected numbers. The reality is that wireless data is a hard sell. With the exception of five key vertical market applications (electrical utilities, insurance, petroleum, public safety, and trucking), wireless data has not lived up to its expectations in the general business or consumer markets.

One issue is performance. Most industry observers agree that the cellular system was designed mainly for voice, not data. For wireless data to succeed, carriers are going to have to keep pace with wired data services. As one industry analyst puts it, "Until you can give consumers on their phones what they get on their computers, I don't see them buying wireless data service."

Another problem is the scarcity of useful data-oriented products that operate seamlessly in a mobile environment. Also, few large software companies have stepped up to support wireless data. Then there are issues such as network coverage, tariff structures, distribution channels, and the actual cost of wireless data services.

All of this is expected to change as traveling professionals make more and better use of their laptops with wireless data products, such as so-called smart phones with integrated modems. The introduction of more powerful, full-function notebook computers, the improved availability of cellular data services, and the growth of end-user applications will also help.

Newton, Massachusetts-based Business Research Group believes that by the end of 1999, 93 percent of medium and large corporations will offer remote access to their employees. That process has already begun.

Figure 4-1. The corresponding growth of use of the Internet and mobile telephones.

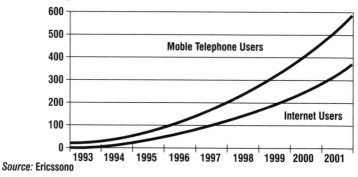

Source: Ericssono

As Figure 4-1 shows, the increasing popularity of the Internet correlates with the growth of mobile data communications.

Currently Available Wireless Data Services

Circuit-switched data is already available across existing analog and digital cellular networks worldwide. Packet-switched data services are also available over dedicated frequency bands through BellSouth Wireless Data (formerly known as RAM Mobile Data before BellSouth acquired it) and ARDIS in several countries over cellular digital packet data (CDPD) networks.

Other standards (e.g., pACT and iDEN) are pushing into the wireless data arena. Also, a number of telecom companies have demonstrated circuit-switched CDPD, allowing cellular operators to cost-effectively offer data service where voice services already exist, mainly in rural or other outlying areas where CDPD has not yet become available.

Most U.S. carriers have already adopted CDPD and some carriers, like Ameritech and Bell Atlantic NYNEX, have linked their CDPD networks to offer seamless roaming coverage throughout their service areas. Canada, Mexico, and New Zealand, along with several other countries, are also rolling out CDPD systems. In Europe, the Global System for Mobile Communications (GSM) should continue to dominate as the data transmission carrier standard. Asian countries will likely adopt a variety of wireless data technologies.

As with anything else, you usually must figure out your requirements before committing to a network technology. Wireless technology choices include the following:

- Spread spectrum can have a maximum wide area range of up to thirty miles between radios, but that drops to 1,200 feet or less in in-building systems, such as in wireless local area networks (WLANs) accessed through laptop computers equipped with WLAN cards.

• Satellites, whether in geostationary earth orbit (GEO) or medium- or low-earth orbit (MEO and LEO), offer global data coverage with data rates ranging from 64 kilobits per second (Kbps) to 45,000 Kbps.

• Microwave communications, including analog and digital point-to-point communications, can replace leased lines in dedicated networks with wide area coverage up to fifty miles.

• Paging is now available with one- or two-way service and 100 percent coverage in the United States and most other countries. Service is available from a variety of carriers, including certain FM broadcasters and mobile communications satellite services.

• Integrated Digital Enhanced Network, better known as iDEN, essentially boosts the performance of cellular networks to combine voice, dispatch, and short messaging with data. Developed by Motorola, iDEN covers most U.S. metro areas.

• DataTAC is a two-way packet radio service that connects TCP/IP (the Internet protocol suite) or X.25 (packet-switching network) protocols and supports transparent and seamless roaming. Also developed by Motorola, DataTAC is managed by the Worldwide Data Networks Operators Group.

• Mobitex is a two-way, packet-data-only network. Developed by Ericsson, it supports broadcast, store-and-forward messaging, and host or peer routing.

User equipment costs vary widely in each of these systems, from hundreds of dollars for a digital PCS or digital-ready cellular phone, to at least $2,500 for a mobile satellite terminal. Service charges may vary just as widely.

Business Applications for Mobile Data

Carriers' data expertise and range of services are making it easier for corporations to extend the use of their information technologies using the Internet or company intranets, and companies are responding with all kinds of applications. Some are obvious, some not so obvious.

For example, more than 3,000 Progressive Insurance claims adjusters are using portable wireless data terminals to settle claims and write checks at the site of an accident. Other growing applications for wireless data include telemetry (e.g., vending machines, meter reading, tracking stolen cars) and credit-card verification, particularly in underdeveloped and developing countries, where there are relatively few phone lines.

Most surveys indicate that more than half of the companies currently using wireless data in their businesses report that the technology pays for itself within a year by improving efficiency in the workplace. In some cases, it pays for itself within six months.

Package-Tracking Systems

In 1992, United Parcel Service (UPS) began developing a nationwide, real-time package-tracking system that combines UPSnet, the company's existing wire-based network, with cellular technology. To provide the service, UPS had to tie together a network of more than seventy large and small cellular carriers. The carriers got on board by providing UPS with a single point of billing for their cellular airtime.

The UPS system (and its component hardware) works like this:

1. When delivering a package, UPS drivers use a small terminal called a DIAD (Delivery Information Acquisition Device) to scan the bar codes on the package's label.

2. When drivers return to the truck, they insert the DIAD into the DIAD Vehicle Adapter (DVA).

3. The DVA transfers the package information from the DIAD and transmits it to the nearest cellular telephone tower via an in-vehicle cellular modem.

4. The data is then routed through the cellular system to the wireline UPSnet, and on to the UPS Data Center in New Jersey.

5. At this point, the package information is stored in a database where it can be accessed by a UPS customer service representative. On an average day, UPS will track well over six million packages using this system.

FedEx has its own remote parcel-tracking system and has added a relatively new wrinkle to its system: Instead of sending trucks out on a regular basis to check on the contents of its widely disbursed drop-boxes (some of which may be empty and therefore not worth the trip), FedEx has equipped some of these boxes with small, solar-powered radio transmitters to signal a dispatch center when something has actually been dropped into the box.

Repair Services

Sears, Roebuck and Co. has implemented a large-scale system for 7,000 of its repair service technicians. Varying by district, depending on coverage capabilities, mobile workers use one of the available national data networks to interface with the Sears IBM 3090 mainframe. With this system, Sears can remain in constant contact with its mobile workforce. Throughout the day, new jobs are dispatched to the technicians, who are able to respond with job status, request price estimates, look up parts availability, and place orders over a wireless network. In the evening, the technicians plug their handheld PC into a phone jack and during the night the unit automatically dials up the host computer. Through this wired connection, the technicians receive automatic software updates and their job schedule for the following day.

Restaurants

To help speed order processing and improve customer service, some restaurants are implementing wireless technologies that allow waiters and waitresses to send customer orders directly from the table to the kitchen or the bar. One such system uses Fujitsu palmtop computers with PC car radio modems to call up food and drink menus, place orders, and process credit cards—all from the customer's table.

In fact, some restaurants have managed to reduce their waiter/waitress staffs with systems that allow customers to place their own orders from their tables without restaurant staff assistance. The experience is not much different from playing an easy-to-use video game.

Car Rentals

Many people who rent cars are familiar with the portable terminals carried by Hertz and other car-rental personnel that eliminate standing in line when returning cars at an airport. Those systems are also now being used in gas stations (by early 1998, Mobil Oil had 3,000 stations across the United States equipped with these systems).

Wireless Credit Card Transactions

First Data Merchant Services, the world's largest provider of merchant processing services for VISA, MasterCard, and other major credit cards, has developed a wireless credit card transaction system using CDPD technology that enables merchants to process bank card transactions from locations where telephone lines are inaccessible or impractical. The system works like standard transaction-processing methods; however, instead of using phone lines for data transmission, the encrypted data is transmitted via a cellular network. The system works particularly well for mobile merchants who operate taxi, pizza delivery, and towing services, as well as for construction-related services and temporary locations (e.g., special events such as air shows and fairs).

Public Safety Networks

Law enforcement has also taken to wireless data. For example, each of the seventeen vehicles in the Bridgewater, New Jersey police department is equipped with a dashboard-mounted IBM laptop computer linked to Bell Atlantic Mobile's CDPD network. The system allows officers on patrol to check license plate numbers without having to call them in to a police dispatcher. Detailed descriptions of wanted criminals or stolen cars can also be transmitted and stored in the system so that the officer doesn't have to write down the information.

Some of the features of this system can be adopted by businesses where security is a critical issue. For instance, the system allows the shift commander to communicate discreetly with one or more officers in the field, which can be a tactical advantage if the police believe their calls are being monitored by a scanner.

Remote Monitoring

BellSouth Cellular Corp. has heavily promoted its Cellemetry Data Service, a nationwide network that can monitor just about anything, including utility meters and vending machines. Cellemetry is also being used as a key element in cellular-based security systems and to remotely monitor railroad-crossing signal equipment.

New Product Development: Hardware and Software

Apple Computer, Motorola, and IBM Corp., among a few others, have come up with some portable devices that, for a variety of reasons, have not been very successful. Some were in the smart phone category; others were personal digital assistants (PDAs). The best (or perhaps worst) example of a PDA is the early version of the Apple Newton.

There are also software programs designed specifically for mobile wireless data. One of the early examples was Ericsson's DC 23 Mobile Office, a hardware and Windows-based software package for sending and receiving faxes and transferring files. DC 23 uses data-compression techniques and supports data transfer at speeds that will accommodate most GSM networks.

Some of the newer hardware and software products and services for wireless data users are discussed in the following sections.

AT&T PocketNet

Web sites have also been created specifically to accommodate cellular and other mobile phone users. One, called Travelocity, is available from AT&T's PocketNet Internet service. It combines an analog cellular phone with a circuit-cellular modem and a CDPD modem and allows travelers to check on flights at the touch of a few buttons. The service is also available on certain types of pagers. One bonus feature is that if you notify the Travelocity site of a flight you are scheduled to travel on, the site will page you if the flight is delayed.

Low-cost, easy-to-use Internet-based dispatch management software has also been developed for business users. Called PocketPower, this package features immediate two-way wireless communications via AT&T's PocketNet telephone and other smart phone devices. It is designed for small to medium-size businesses that repair, deliver, or install products and require immediate dispatch of their field service technicians or vehicles.

PDA/Smart Phone Combinations

One of the hottest products for wireless data applications is Nokia's 9000i Communicator. At first glance, it looks like most cellular phones, but it also functions as a handheld personal computer. The 9000i can send and receive e-mail, faxes, and other short text messages, as well as browse the Internet, corporate intranets, and public databases.

Several new pagers are also wireless data enabled, with some two-way models featuring a full keyboard and software with web browser capability.

Ricochet

One wireless service that is dedicated to data is Metricom's Ricochet, available in the San Francisco Bay Area, Seattle, Washington, D.C., and several airports and college campuses. With Ricochet, users can go anywhere in their service area and, using a modem-equipped laptop or notebook computer, send or receive e-mail, access the Internet, check local movie schedules, or browse online versions of *The Wall Street Journal* or *The New York Times.*

Broadband Wireless Data: LMDS

More commonly, however, carriers offer mobile data in conjunction with voice services.

A new wireless data alternative suitable for fixed rather than mobile use is the local multipoint distribution system (LMDS).

Originally developed by CellularVision for highly localized cable television services, LMDS has evolved into an advanced wireless broadband network for data transmission. The system uses high-frequency microwaves in the 28 GHz range to send and receive broadband signals over short distances. The cells are relatively small, covering a radius of two to three kilometers from a single hub transceiver.

LMDS is suitable for voice and video, as well as data. Although the FCC auctions for LMDS spectrum were initially disappointing, LMDS is expected to begin making fairly significant inroads in the business market by the year 2001.

Infrared Connections

An alternative to radio frequency (RF) for data transmission is infrared technology. Most people know infrared since it is the technology used in the TV remote control that is popular in so many homes. Infrared (IR) uses light to carry information, which means that it works only if there is a line-of-sight connection between the transmitter and receiver. Another limitation to IR is that misty environments will cause transmission problems.

The Infrared Data Association (IrDA), together with the dominant market players in the cellular and paging industries, have put together a set of standards that will empower wireless communications devices, such as cellular phones, pagers, and personal computers, to transfer information over short distances using IrDA-based IR data communication ports. The IrDA specification embraces established IrDA protocols and other industry standards to advance applications that will transfer electronic business cards, calendar events, telephone book entries, messages, real-time audio, and other information seamlessly between multiple brand-name digital devices.

The adoption of the standard is expected to lead to the appearance of IrDA ports in a variety of electronic devices, including 3Com Corp.'s new third-generation Palm III electronic organizer. The Palm III can send and receive data wirelessly via a modem attachment, or it can transfer data back and forth with an appropriately equipped desktop computer using an infrared interface.

The Killer Application: Wireless Internet Access

As this chapter has shown, there are many specialized or vertical market uses for wireless data products and services. But it is increasingly apparent that if there is a "killer application" for wireless data that crosses over to the mainstream or horizontal market, it is the Internet. Traffic on the Internet doubled every 100 days through 1997 and the early months of

1998. In fact, the Internet tops any technology that has come before it in terms of rate of growth. Radio, for example, was available for thirty-eight years before it had 50 million listeners, and television took thirteen years to reach that level. The Internet hit the 50 million mark in just four years.

As more people come to rely on the Internet for work and personal use, they will expect access to the Internet even when they are on the move. Industry analysts are projecting the American market alone for equipment, carrier services, and software providing wireless Internet access for both fixed and mobile wireless applications will climb to $37 billion by 2002, from less than $3 billion in 1997.

Wireless access devices for the Internet will range from portable digital phones (such as AT&T's PocketNet) to laptop and notebook computers. There is also the fast-emerging category of new handheld PCs. Networks, too—including the public cellular network, PCS, wireless local area networks, mobile satellite services, as well as dedicated data networks such as ARDIS—will likely benefit from the growing popularity of wireless access to the Internet and intranet services.

CHAPTER 5

Mobile Computing— You Can Take It With You

That tiny antenna sticking out of the handheld personal computer (HPC) isn't there simply to impress anyone. It actually lets HPC users accomplish most of the work they normally do at their desktop PCs, including accessing corporate databases and public and private networks.

Increasingly, "portable computer" has come to mean "mobile computing" as a growing legion of business users has adopted a variety of products and services that provide never-before-possible mobility. You no longer have to be sitting at your desktop to check your e-mail, surf the Web, word process, crunch numbers, or do any of the other things that you used to do only at your desk.

For a growing number of companies, this has become an unexpected bonus in their efforts to improve productivity. One-third of the U.S. workforce, or 43 million, is mobile—that is, working 20 percent of the time away from the primary workplace. More companies are pushing their workers out into the field and abolishing fixed office space, driven by demand to stay competitive. These mobile workers need easy access to huge resources of enterprise data in addition to simple messaging capabilities.

Internet/Intranet Trends:
A Double-Edged Sword

Clearly, the Internet and intranets have dramatically changed the way companies conduct business, extending the network's usefulness out to the remote and mobile enterprise. According to a survey taken by The Yankee Group, 87 percent of the top U.S. companies were developing intranet applications by the end of 1997. The pervasiveness of intranets is expected to stimulate efforts to incorporate remote workers into integrated business practices. However, among corporations today, the focus is more on using intranets to connect remote company locations than on providing mobile access.

The Yankee Group found that only 20 percent of the companies it surveyed provided remote Internet/intranet access via wireless to their mobile workers. That is likely to change, and fairly rapidly, despite a number of issues that corporations and users are still wrestling with.

For one thing, the World Wide Web (WWW) has created an expectation, in terms of information delivery, that cannot be met by wireless data networking—at least, not yet. At the same time, the attributes of the Internet can drastically reduce the cost and complexity of wireless data integration. By cascading the two technologies, it may be possible to break the logjam that has stifled wireless mobile data in the form of underutilized networks, low revenue, and lack of applications.

Portable End-User Devices

What often confuses the uninitiated is the several categories of computing devices to choose from, and the fact that just about all of them can be configured to communicate untethered by wires. These products include laptop computers equipped with a PC card modem, which most people are at least somewhat familiar with, to a smaller class of devices—HPCs, or notebook computers; subnotebook computers; the hugely popular

Figure 5-1. Siemens' FoneBook PLUS.

PalmPilot, personal digital assistants (PDAs), some of which may fall into the slightly less elegant electronic organizer category, and so-called smart phones, probably the best example of which is the Nokia 9000i Communicator, a clamshell-like device that looks like a chubby cellular phone, but which opens up into a full (but tiny) QWERTY keyboard with easy-to-read display and impressive functionality. Another smart phone example is the Samsung phone used in the AT&T PocketNet system, which provides basic Internet access. Siemens has come out with FoneBook PLUS, which enables users to store, change, and download names and numbers from a PC to the Siemens g1050 PCS phones (see Figure 5-1).

The mobile computing device most people choose is usually based on its usefulness—in other words, its functionality—how well it fits their needs, and, to a lesser extent, its form factor (size, weight, size of the keyboard and display).

The best bet currently for general business use may be HPCs, virtually all of which offer wireless modems.

HPCs and Microsoft Windows CE

One of the most important developments in mobile computing is the HPC. Several of the biggest personal computer manufacturers have introduced HPCs that operate under Windows CE, which is a somewhat scaled down version of Microsoft's Windows 95. The current version, Windows CE 2.0, is significantly improved over the original, but still has a way to go, according to the more sophisticated users of CE-based products. This should not be a long time in coming because Microsoft has not only opened up Windows CE to a growing number of microprocessors, but also hopes to get Windows CE into smart phones, office automation, and other telecommunications to function mostly as embedded controllers.

HPCs with Windows CE offer instant synchronization with desktop PCs or laptop computers via the serial port and direct-connection cables or cradles. Casio's Cassiopeia HPCs can even download pictures from the Internet into the Casio QV-300 digital camera through a PC card and can print out a digital picture from the camera.

More than one million PalmPilots were sold in 1997; at least nine HPC vendors like the number enough to announce plans to introduce at least slightly different versions of the PalmPilot by the end of 1998. These will most likely be Windows CE-based devices. It is also likely that the popularity of the PalmPilot and Nokia's 9000i will create more interest in smart phones that can both communicate and commute in a mobile environment. In fact, if smart phones become powerful enough, they could eventually replace notebook computers for mobile professionals.

Also, several wireless messaging service providers have signed on to Windows CE for future HPC-enabled wireless communications services, and a number of independent software developers are working on applications for Windows CE-based devices.

GTE, for example, has combined its PageCard wireless messaging system with Socket Communications, Inc.'s Page-Card data pager and PageSoft software to enhance HPCs using Windows CE. Users will be able to receive "paged" messages directly into the Windows CE PMAIL Inbox messaging tool via the PageCard. The SkyTel Messenger for Windows CE operates with SkyTel 2-Way devices, or with Windows CE-based HPCs.

AirMedia offers a full range of direct-to-the-palmtop wireless Internet broadcasting services for Microsoft's Windows CE Version 2.0. AirMedia Live is the company's wireless network for broadcasting public interest news from more than sixty sources, including CNN, Reuters, CBS Sportsline, UPI, Forbes, The Weather Channel, and the BusinessWire. It also accesses information from corporate intranets and existing corporate IT systems, as well as personal alerts to announce the arrival of new e-mail on the corporate mail server, the confirmation of a trade, or complete Internet wireless messaging from public and corporate e-mail and Web sites. By wirelessly delivering the Internet to mobile computing devices, AirMedia Live bypasses the need for an Internet connection or scheduled content downloads. It extends the reach of the Internet to connected and nonconnected users without tying up a phone line. In order to wirelessly enable a wide range of products, AirMedia licenses its technology to device manufacturers such as Philips and Compaq for Windows CE.

NEC ships its Windows CE Manager software compact disk for installation on its desktop or notebook computers. ARDIS, a wireless data network, supports the Philips Mobile Computing Group's Velo 1 HPC and the MobilePro with Windows CE. AT&T Wireless Services has demonstrated Windows CE with Compaq's HPC and a Motorola modem on cellular digital packet data (CDPD), a service that uses idle time in the analog cellular phone system to transmit packet-size data at rates up to 19.2 kilobits per second.

Uniden Corp., meanwhile, is using its UniData 1000 wireless CDMA modem and the wireless service of GeoAmerica Communications Corp. to provide Windows CE-based wireless access to the World Wide Web over the CDPD network. GOAmerica runs an application called AirBrowse, which allows users to run standard Web browsers such as Netscape or Microsoft Explorer and to wirelessly enable these browsers when the user is mobile. RadioMail has also demonstrated a multinetwork wireless Internet-access application for Windows CE for use with ARDIS, RAM Mobile Data (which operates the Mobitex network throughout the United States and an ARDIS competitor), CDPD, PCS, and the European all-digital Global System for Mobile Communications (GSM) cellular system.

The RadioMail system supports wireless workgroup messaging between members of the group and the mobile staff with paging, faxing, access to stock quotes and news, and access to any site on the World Wide Web.

Even though Windows CE was launched for handheld devices, Microsoft believes that telephones, television sets, and automobiles will expand the deployment of Windows CE for use in personal information managers—allowing, for example, business travelers to have instant access to their schedules and client lists and to be able to match this information to data in a laptop computer via an infrared port.

The PalmPilot gained popularity as a device that can organize and track business information and memos, but newer models can send and receive e-mail and access the Web wirelessly with a tiny snap-on modem. The modem adds slightly to the PalmPilot's size and weight, but also adds significantly to its functionality—that is, as long as the user is in an area with cellular digital packet data (CDPD) wireless data coverage. Unfortunately, some U.S. cities still do not offer CDPD coverage.

All of the currently available HPCs are equipped with Type II PC card slots and an infrared (IR) interface, and many HPCs and some smart phones will eventually feature Global Positioning System (GPS)-based navigation and E-911 position location capability. GPS will either be integrated into HPCs or phones or added through the use of a Type II PC card module.

Actual Use Scenarios

The bad news for many users is ease of use. It usually takes a dedicated road warrior to sit down at an airport, for example, and begin plugging everything together, download e-mail messages (especially during a busy Internet period), and respond, then disconnect everything and pack it up.

Several options are available, particularly to laptop users. These include subscribing to one of the public wireless data networks—BellSouth Wireless Data (formerly RAM Mobile Data), ARDIS, Metricom's Ricochet network, or CDPD. Each of these has its own features and benefits that would lend themselves to a specific requirement, and all of them will continue to be upgraded over time to meet new and emerging user needs. (Metricom, for example, has enhanced its Ricochet service with a new pocket-sized modem with longer battery life, a desktop modem for fixed-location use, and a new dial-in access service for out-of-coverage area use.)

Metricom's Ricochet, while fast and portable, is available only in a few locations, including Seattle, the San Francisco Bay area, and Washington, D.C., as well as some college campuses. Of course, that will change over time.

CDPD is available on a much wider scale, but also has a limited nationwide service and requires a roaming fee outside Bell Atlantic's home territory. ARDIS and BellSouth Wireless Data (RAM Mobile) provide coverage throughout the United States, but are a bit slow when transmitting long messages.

Because of these current limitations of ease of use and coverage, Windows CE-equipped HPCs are (for the moment) used to perform more mundane tasks.

Hoechst Marion Roussel (HMR), an international pharmaceuticals manufacturer, for example, has equipped its North American sales force with HPCs. For almost ten years, the Food and Drug Administration (FDA) has required physicians to sign for every prescription drug sample they receive. In 1988, HMR developed its own software application for its sales team's pen-based laptop computers to capture the required signatures. In searching for a more efficient mobile computing tool, the information technology staff at HMR found that it could easily apply this custom application to the Windows CE operating system.

In another application, Goldman, Sachs & Co. has automated a part of its trading process by providing its brokers with a custom application that runs on HPCs equipped with wireless modems.

Wireless Thin Clients

With promises to boost productivity and lower the cost of business computing, a growing number of companies are migrating to client/server computing and providing users with their own fully loaded PCs. Although this has empowered PC users, it has also increased maintenance and upgrade costs. While the purchase price of a PC may be relatively inexpensive, the cost of deploying and managing a computer for each user has exploded. In fact, the Gartner Group estimates the cost of owning one desktop PC at up to $12,000 a year, much of which it attributes to upgrades, maintenance, and support costs.

Information technology (IT) managers and their companies want to reduce their computing costs. At the same time, the need to support standardized and consistent configurations, maximize the useful life of their technology investments, and provide secure computing environments has become both a short- and long-term goal.

Characteristics of Thin-Client/Server Computing

As a result, a new category of wireless devices, known as thin-client terminals, has been developed for the millions of locally mobile workers—professionals who spend much of their workday away from their desks but within a building or campus, and who need to access computer resources while moving around. They require productivity devices for communications that are truly mobile. Thin-client terminals provide little or no local storage or processing and rely on the network for applications, data, and functionality. Essentially, they work by combining wireless connectivity with corporate networks to access the applications and resources of a remote server or a computer mainframe.

Simply, wireless thin clients allow locally mobile workers to access standard Windows, Windows NT, and terminal-based applications anywhere in the workplace with a user-friendly "electronic tablet" that is wirelessly linked to a Pentium or Pentium Pro server. A single server can host more than sixty thin clients.

What is new here is that thin clients have little or no local storage or processing, so they rely on the network for applications, data, intelligence, and functionality. Thin clients are streamlined devices designed primarily to access the applications and resources of a remote server or mainframe that do not incur the processor, memory, or storage costs of an independently functional PC.

This new computing architecture can be divided into three parts:

1. The mobile wireless thin-client device

2. A Windows NT server

3. The radio frequency (RF) network that joins the two

Users interact with a wireless thin client just as they would with a PC, except that they touch the screen with a finger or stylus rather than clicking a mouse and use an on-screen keyboard instead of a desktop keyboard. The input (keystrokes and mouse clicks) is sent across the RF network through an access point (AP)—essentially, a networked radio that bridges wireless local area network coverage onto the wired Ethernet backbone—to the wireless thin client. There data are displayed in real time, making the performance and response time of this system (with the exception of motion video or graphical gaming applications) pretty much on a par with the performance of a desktop PC.

With thin clients, everything but the display executes on the server, enabling users to take advantage of server-class processors, memory, hard drives, and network connections. The servers can be upgraded, which means that the portable device itself will not be made obsolete by next-generation processors. Users can also add new memory or applications by upgrading only the server, with changes reflected in all the thin-client devices. Standard Windows NT tools manage maintenance and support.

Thin clients also optimize wireless bandwidth. Traditional "fat PC" client/server architectures use a large amount of bandwidth in unpredictable patterns by requiring the user to upload and download various data files to and from the server. Wireless thin-client devices minimize network traffic by transmitting only input—keystrokes and mouse clicks— and screen refreshes. And since data reside permanently on the server, and never on the client device, thin clients provide the data security of a "mobile monitor." If they are lost or stolen, the central processing unit (CPU), hard drive, and all of the data remain centralized on the server.

Mobile Network Computers

Mobile network computers (NCs), on the other hand, download Sun Microsystems' popular Java operating system applets (usually small, personalized applications) and then execute these applications locally. The key to NCs, which are supported by several name-brand computer and telecom vendors, is that they allow you to plug into the network in your office or into anyone else's office network.

It might work like this: You're in the office, cleaning up a few last-minute details before a business trip. Before departing, you plug your mobile network computer (NC) into your company's local area network to download some information from the company's intranet site. On the way to the airport, you connect to the Internet using your cellular phone to download your e-mail. While in the airplane, you read and reply to the e-mail and peruse the information that was previously downloaded from the intranet. Your mobile NC has enough intelligence to know that it is in a "disconnected" mode and waits until the next connection to send the e-mail that you replied to while on the plane. You do that by simply plugging the mobile NC into your client's (or any other) local area network.

Most of this functionality is available today. The hard part is plugging into any local area network to access your own local area network. That should change with the adoption of something called the Mobile NC Reference Specification (MNCRS), the key to which is the widespread adoption of the Java operating system. MNCRS/Java will give software developers a common reference platform on which to write their NC applications.

Products that would fall under the domain of the MNCRS would be mobile network computers and "smart" cellular phones, as well as virtually any new category of lightweight, handheld data device. The common thread is that they all would offer easy access to the Internet or corporate networks. The MNCRS standard is designed to determine how mobile computer screens should look, their power requirements, how they would be linked with networks, and the types of peripheral devices they would support.

Significantly, no minimum bandwidth is specified in the MNCRS. Any MNCRS-compatible device will communicate at various connection speeds, depending on the transmission medium.

For example, when connected in the office, the unit would typically talk over a high-bandwidth connection. When the user leaves the office and makes a wireless connection, any information, including e-mail messages, that does not require a high-bandwidth connection could be downloaded.

Comparing NCs and Thin Clients

Thin clients are gaining attention for the locally mobile worker—initially in vertical markets such as healthcare, manufacturing, retail, and vehicle service. These people work while moving around within a building or campus, rather than sitting at a desk.

The primary difference between the thin-client/Windows-based terminal and an NC is that thin clients relegate all application functions to the server, projecting only the presentation layer or user interface to the client device. NCs, on the other hand, download what are essentially custom Java applications and then execute those applications locally. Because thin clients provide centralized management of all computing resources and can run existing Windows or terminal applications, its proponents believe they have the edge, at least for the time being.

Industry Examples of Mobile Applications

The Internet-enabled CruiseConnect technology developed by Cruise Technologies, for example, would allow a medical organization with headquarters in Chicago and branch office clinics in other cities to easily centralize all computers in Chicago. The other clinics, without on-site computing, could access and update any of that information in real time, using mobile thin-client devices that communicate wirelessly in the branch office, across the wired Internet, to the computers in Chicago.

Motorola's Worldwide Data Solutions Division is using its SidePad wireless thin client terminal with Proxim, Inc.'s wireless local area network APs in manufacturing and warehousing applications.

In a manufacturing environment, the local-mobile concepts provide managers and workers on the manufacturing floor with instantaneous access to accurate information. Warehouses could connect to inventory records and purchase orders, thereby reducing the risk and inefficiencies of transferring numbers through paper forms.

In healthcare, users have access to complete and accurate patient data gathered and displayed at the point of care. Vehicle maintenance facilities could use this technology to automate the inspection and repair process. Retailers would have instant access to pricing information, store inventory, or real-time sales affecting decision making by managers.

However, Federal Express has replaced thousands of terminals and PCs with NCs at its Memphis headquarters and its branch offices. Rather than supporting new or upgraded PCs, FedEx believes that it can save up to $250 million a year by shifting to NCs.

Others are introducing thin-client terminals with a PC card slot that would allow the terminals to be used over CDPD-based digital transmission networks.

Otis Elevator, with a service organization of more than 1,900 people, uses the ARDIS network to dispatch approximately 300,000 service calls a month. Otis dispatches calls, orders parts, and updates customer maintenance records via its application over the ARDIS network.

The key in all these cases is mobility. Applications that were once too costly to do remotely, using devices that were too power hungry or too bulky, have become very practical—to the extent that they are paying their own way. As time goes on, these devices will continue to be improved with upgraded functionality and ease of use.

CHAPTER 6

Wireless Telecom Regulations—Confusion and Competition

R ecent changes in the way telecommunications services are regulated, both in the United States and
worldwide, have had a dizzying effect on wireless
communications services. The most dramatic—and most
contentious—event has been the passage of the Telecommunications Act of 1996. With the possible exception of
the consent decree signed off by the Department of Justice
and AT&T in 1982 (which, among other things, opened up
the market to competition in long distance service), the
Telecom Act represents the first major change in telecommunications law since the passage of the Communications
Act of 1934.

The Telecom Act was supposed to throw open the doors to
telephone, cable, and other companies eager to provide information and communications services. The idea was to write
new laws that would promote competition in local exchange
areas and increase competition in the long-distance market.
In fact, the new law deliberately blurs the lines between formerly discrete sectors of the telecommunications industry. It
was also supposed to result in lower prices for both business

users and consumers. So far, this has not happened, at least not in any significant way.

The vagueness of the new law, which has brought on several court challenges, along with rapidly changing technologies and the realities of running a highly complex and fast-changing telecom business, has forced the industry to scramble to try to develop a coherent strategy for the future.

LATA Confusion

The Telecom Act goes well beyond the consent decree in that it seeks to preempt state and local laws from barring competition in state and local markets. It also allows regional Bell operating companies to offer interLATA telecom services (i.e., services that originate in one and terminate in another local access and transport area, or LATA) in their regions if they meet certain requirements.

The Telecom Act allows the former Bell companies to immediately offer long-distance and international services outside their operating territories and to their cellular customers. But these companies can provide in-region long-distance service only after their local service markets are open to competition. It also opens up the possibility that AT&T or MCI may become your local telephone company—or your source for myriad wireless services. And cable system operators such as Cox Communications or Comcast may offer broadband Internet or wireless local loop services.

Intent of the Telecommunications Act of 1996

Under the new law, each Bell company must meet a fourteen-point checklist before the Federal Communications Commission (FCC). After consideration of any input from the Department of Justice and the relevant state regulatory authorities, the FCC can grant these companies permission to offer long-distance services to their local customer base.

The checklist requires the Bell companies to provide competitors with a link to their networks and to give competitors access to individual network elements and other "unbundled" elements. Competitors include cable television and long-distance companies such as AT&T and MCI. Competitors also may simply resell Bell local service; state regulatory commissions have set resale prices that discount existing service prices from 17 percent to 25 percent.

The Telecom Act further allows the Bells and other local exchange carriers to own cable TV operations, although this ownership is limited to not more than a 10 percent financial interest in an existing cable operator within the Bell's own territory. Once a Bell company has approval to offer in-region long-distance service, it also may begin manufacturing telecommunications equipment. The Telecom Act also enables GTE, which offers local service in parts of twenty-eight states, to immediately enter the long-distance business. The Telecom Act even allows utility holding companies to provide telecom services.

Unfortunately, like a lot of legislation, the Telecom Act is unclear on some issues, such as whether the regional Bell operating companies need to have separate subsidiaries for personal communications services (PCS), as is the case for cellular. The Act does, however, allow cellular and PCS carriers to pursue residential wireless local loop (WLL) services.

Reality Check: More Costs, Less Competition

One of the problems is the way the Telecom Act evolved. Lawmakers who supported the new Telecom Act misjudged the pace at which the cable industry and wireless local loop technologies would begin to compete with conventional local wireline services.

Cable companies, after promising to bring more competition (and presumably lower prices) to the telephone market, have backed off in the face of unanticipated technical difficulties and a change of heart about investing in new facilities. Wireless local loop services, although still very promising, have been slow to develop as a viable competitor to the traditional wireless carriers.

While the Telecom Act creates a framework for changes in the telecom laws, it left the writing of the actual rules to the FCC and the states, which were still hammering out the details of hundreds of issues well into 1998. With many of the new rules unacceptable to key segments of the telecom community, the Act has been challenged in several lawsuits and went into virtual legal limbo when a U.S. District Court struck down as unconstitutional a provision that prevents the Bell telephone monopolies from selling long-distance service in their own territories. (Under the Act, the Bells are allowed to sell these services only after they have opened up their local telephone markets to long-distance companies and other competitors.)

The result of all this is that more than two years since the passage of the Telecom Act, there is more confusion—and not much more actual competition—in an industry that has been unwilling to invest aggressively in the development of new services until the political and legal dust settles.

Market Breakdown

According to *U.S. Industry and Trade Outlook 1998: Telecommunications Services,* a report by the U.S. Commerce Department, the Bell operating companies and more than 1,000 independent telephone companies in the United States continue to be the sole providers of local telephone services in their franchised areas.

While well over 90 percent of wireline service revenues are controlled by about a dozen large companies (the seven former regional Bell companies, GTE, and the four largest long-distance companies), the introduction of new services based on new or enhanced technologies provides ample opportunities for niche market players in telecommunications services.

About 130 firms offering long-distance services do so over network facilities of which they own at least a part. Another 300 to 400 companies provide toll services by reselling the long-distance services of the large carriers (e.g., AT&T, MCI, and Sprint) that have built their own nationwide networks. Sixty or more competitive access providers (CAPs) have built fiber networks in urban business districts that connect with the networks of long-distance carriers.

On the wireless side, about 950 firms provide cellular, paging, and other mobile services, although this number is shrinking with acquisitions, mergers, and consolidations.

Impact on Wireless

What does the Telecom Act mean to wireless communications service providers and users?

A study by A. T. Kearney, an international management consulting subsidiary of Electronic Data Systems (EDS), says the law provides a competitive advantage to operators of wireless systems once their systems are completely built out. The legislation allows cellular and PCS providers to provide bundled packages of local and long-distance services without offering the presubscription obligations that incumbent landline carriers will be required to offer.

However, the FCC may require unblocked access to long-distance carriers through access codes or 800 numbers if it determines that customers are being denied the use of long-distance services of their choice and that this denial is contrary to the public interest. Indeed, at least two regional Bell operating companies—Southwestern Bell Wireless and Ameritech—entered the cellular long-distance market immediately after the Telecom Act was passed. Others will follow.

Two key provisions of the Telecom Act that apply specifically to the wireless industry address siting of wireless facilities and a wireless equal access requirement.

1. *Wireless facilities.* Although state and local governments retain their authority over zoning and land-use matters, these local jurisdictions are prohibited from passing or enforcing ordinances that would make it impossible for wireless providers to erect necessary facilities. Local jurisdictions must act on siting requests in a reasonable period of time.

The Act also specifically prohibits state and local governments from basing site regulations on the environmental effects of radio frequency emissions if those facilities comply with the FCC's regulations concerning such emissions. Finally, the legislation directs the President to prescribe procedures for making federal property available for wireless telecommunications infrastructure sites.

2. *Equal access.* Wireless providers are also not required to provide equal access to common carriers for telephone toll services. However, if the FCC determines that customers are denied access to the telephone toll service provider of their choice and that such denial is contrary to the public interest, the FCC can prescribe regulations to afford customers unblocked access to the provider of the subscriber's choice through the use of a carrier identification code or other mechanism.

Another change is the way the Bells do business. Before the Telecom Act was passed, the Bell companies' cellular subscribers were required by law to select a separate long-distance carrier when they signed up for local cellular service and were billed for the services separately. Now, Bell customers receive one monthly bill for both local and long-distance cellular service from one service provider.

Making a Market: New Opportunities Created by the Telecom Act

In effect, the Telecom Act also creates a market for terrestrial microwave systems as network operators turn more to wireless solutions for new services. For example, several competitive access providers have deployed WinStar Telecommunications' 38 GHz wireless local loop bypass technology, sometimes referred to as wireless fiber.

Using very small antennas mounted at the carrier and at the customer's premises to bypass the local exchange carrier's network, traffic is transmitted over a multichannel millimeter wave link with a range of at least five miles. Several regional telephone service providers plan to use this technology to alleviate congestion in their networks caused by increased demand for Internet access.

Regulatory developments will also create future markets for terrestrial microwave equipment and services, such as intelligent transportation systems (ITS) and high-speed wireless data links.

• *Intelligent transportation systems.* In December 1995, the FCC created a new millimeter wave service, assigning frequency bands for smart call systems, such as vehicle radar, automatic cruise control, and lane guidance systems for ITS applications. These applications were granted exclusive use of this spectrum in order to avoid interference problems between vehicles and other communications or radio-controlled services.

• *High-speed wireless services.* In late 1996, the FCC announced its intention to establish a new Wireless Communication Service (WCS) in the 2.3 GHz band. And in January 1997, the FCC set aside 300 MHz of spectrum in the 5 GHz band for the Unlicensed National Information Infrastructure (U-NII) for short-range, high-speed wireless digital communications. The U-NII spectrum allocation, which would not require licensing by the user, would support the creation of wireless local area networks for community use, mainly schools and libraries.

Universal Service

A relatively new but important issue in the United States is the concept of universal service, which essentially means providing basic voice telephone service at affordable rates to everyone in the country. Congress and the FCC have implemented the universal service program to subsidize telecom services for rural and low-income communities, schools, libraries, and rural healthcare facilities throughout the nation.

Funding for the program is derived from an assessment on the gross revenues of all telecom service providers, including wireless carriers. Phone bills now feature a line item designated the "Universal Service Assessment." As the size, scope, and administration of the service program is solely within the discretion of the federal government, the charge may be modified at any time, with or without notice.

Wireless systems are expected to play a key role in the development of universal service because they provide terrestrial and satellite-based services with broad coverage and different cost structures. Wireless local loop systems, for example, can provide voice and data services in areas where they are now available in a limited way (mainly in rural areas) or where local telecom service providers are unwilling to cover areas that are growing and need new or enhanced services.

In the future, universal service will also mean two-way data communications capability that would allow subscribers to access the Internet and other online services. Most terrestrial wireless access systems currently allow data to be transmitted at 9,600 bits per second, the speed of a moderately good wireline modem, to access online services and for other applications. Some of the newer systems designed specifically for wireless local loop systems offer even higher fax and data transmission rates. In some rural areas, where copper cables may be deteriorating, fixed wireless systems may be better able to support high-speed fax and data communications.

However, universal service will be available only if regulators provide sufficient spectrum to support these services.

A World View

Universal service is also a global issue. When the European Union's telecom market opened up on January 1, 1998, it marked the first time in an area of 380 million people that competition in telecommunications was the rule—in local, long distance, and global markets. However, some countries, like Denmark, deregulated their telecom industry several years ago with the privatization of their formerly government-owned telecom carriers. In most cases, the move has increased competition as new telecom companies have begun competing for a share of the market.

In fact, the theme of the 1998 *World Telecommunications Development Report* (WTDR) from Geneva-based International Telecommunications Union (ITU) is universal access. In the ITU report, the Independent Commission for Worldwide Telecommunications Development says that its goal is that "virtually the whole of mankind should be brought within easy reach of a telephone" by the early part of the next century.

This may not be such an easy task. According to figures published in the ITU report, the majority of the population of developing countries—60 percent of the total—lives in rural areas. Yet in these countries, more than 80 percent of main telephone lines are in urban areas.

There are startling disparities in the worldwide distribution of new types of networks and services. For example, 84 percent of mobile cellular subscribers, 91 percent of all facsimile machines, and 97 percent of all Internet host computers are found in developed countries.

Among all forms of telecommunications, mobile communications has experienced the largest and most rapid growth in new market entrants. Even so, the ITU report notes that mobile cellular has had mixed success in enhancing universal access in the developing world. Here, the relevant indicator is the substitution rate—the ratio of mobile cellular subscribers to total telephone subscribers—which indicates the degree to which mobile cellular is being used as an alternative rather than a supplement to fixed-line networks.

Substitution typically occurs where relatively low levels of fixed-line density are combined with competitive mobile cellular markets. The availability of cellular services can also serve to enhance access when the fixed-line network has been extensively damaged due to civil unrest or natural disasters.

To date, the ITU says that cost remains the biggest impediment to cellular becoming a viable alternative for first-time telephone users in developing countries. As the ITU points out, this has opened the door to wireless local loop to the construction of traditional wireline networks. Clearly, the new and emerging satellite networks will compete for "new" business in developing countries.

Other approaches for improving access to telecommunications are to bundle voice telephony with other services, such as cable networks, or to offer Internet telephony (voice telephony bundled with Internet access). Although Internet telephony is unlikely to increase the availability of service due to high start-up costs, the ITU believes that it may help to make it more affordable, as low tariff rates put pressure on public telecommunications operators to reduce their own tariffs.

What is clear is that few countries, if any, have achieved universal service. However, with the liberalization of the telecom industry, universal service is getting more attention. At the same time, the rapid increase in international mergers and other strategic alliances is expected to help improve telecom services worldwide.

Worldwide Spectrum Allocation

The real message for improving the information infrastructure is that it may take more time and involve more changes than many experts anticipated. Regulating this industry will continue to challenge national and regional legislators, as well as global regulatory bodies such as the ITU.

One of the more important legal and political mechanisms for regulating the global wireless sector of telecommunications is the World Radiocommunications Conference (WRC), an international forum for developing and regulating the use of radio frequencies and satellite orbits. The WRC meets every two years, usually in Geneva, with the purpose of reaching a consensus on changes in the use of the RF spectrum, as well as setting the stage for future technological developments.

But with so many nations with dedicated and often self-serving interests, the meetings are almost always contentious at best. The FCC, State Department, and other U.S. government agencies are already gearing up for WRC 1999 and are developing several position papers in hopes of winning international support for spectrum use.

The reality is that with so many new services being introduced, there are few options. It comes down to either withdrawing a spectrum allocation from an existing service, or sharing services in the same band.

One of the WRC's most important tasks is to examine and rule on requests for introduction of new services or expansion of existing services. As the usable portion of the frequency spectrum becomes more heavily subscribed, and as more and more new services apply for the allocations needed to make their systems operational, the stakes at each conference are getting higher and higher.

This was particularly noticeable at the WRC 1997 conference where more than twenty operators sought spectrum allocations and orbital slots to launch global mobile personal communications satellite systems.

More than forty agenda items were included in the WRC 1999 agenda even before WRC 1997 was adjourned, prompting the ITU to suggest in a press release that WRC 1999 "promises to be at least as punishing as that of the 1997 conference."

The ITU's Role

What is the ITU? First of all, as an agency of the United Nations, the International Telecommunications Union's most important mission is to sort out private sector and international political policies of the world's telecom community. Given that the ITU has 188 member nations and more than 450 private sector members, this is not an easy task.

The ITU's next big move in the wireless world will be to issue a critical report at the end of 1999 for what it calls International Mobile Telecommunications 2000, or IMT-2000. The report lays down the concepts and some actual technical details for the next generation of portable/mobile phones—at least in Europe. To do this, the ITU has been promoting a consensus-building process in response to pressures from proponents of a number of different technologies. (For more information on IMT-200, see Chapter 7.)

In March 1996, the Europeans launched the Universal Mobile Telecommunications System (UMTS) Forum. UMTS has been moving aggressively to develop its own set of proposals for third-generation portable/mobile communications systems and services. As currently envisioned, UMTS will be designed to support wideband wireless multimedia capabilities over mobile communications networks, including video, Internet/intranet, and high-speed data communications, as well as voice services.

In January 1998, the European Telecommunications Standards Institute (ETSI), the European Community's telecom standards body, announced its decision to accept a single standard for UMTS. ETSI said that it made the choice based on a proposal submitted jointly by several of the world's leading telecom equipment manufacturers, including Alcatel,

Bosch, Ericsson, Italtel, Motorola, Nokia, Siemens, and Sony. While all of these manufacturers say they will promote a single technology for UMTS, they may also, individually and simultaneously, support other technologies, which some of them have done in the past.

The Japanese and Koreans, meanwhile, have worked hard to speed up the ITU's technology evaluation process, with Korea announcing an extensive joint public-private effort to develop its own proposal for submission to the ITU for consideration as a world standard. That proposal will likely be based on code division multiple access (CDMA).

The ITU's technical evaluation of UMTS was scheduled to be completed by the end of 1998.

FCC Lowers Barrier to Foreign Investment

Deregulation has had an extraordinary impact on the growth of the wireless services worldwide. But it has also muddied the near- and long-term outlook for interconnection rates (and, therefore, user fees) and competition, particularly in Europe where disputes continue to rage between monopoly carriers and new telecom market entrants.

In fact, one of the more contentious issues currently facing the FCC is foreign investment in U.S. telephone and satellite operators, although rules adopted in early 1998 have brought the United States into compliance with a landmark World Trade Organization (WTO) decision to open telecommunications markets around the world. Sixty-nine countries signed the WTO pact, including the United States.

Essentially, the WTO agreement covers three areas:

1. Foreign investment in telecom services and facilities

2. Market access and national treatment of suppliers
of telecom services (Specifically, this part of the agreement calls for fifty-two countries to guarantee access
to their markets for international services and
facilities, with five more countries open for selected
international services. Fifty-six countries will open
up for all or selected services provided by satellite
suppliers.)

3. The acceptance of procompetitive regulatory princi-
ples that the WTO believes are essential to ensure that
new firms have access to markets that up to now were
dominated by monopoly suppliers

The Changing Regulatory Landscape

What's next? To be sure, the FCC has more than it can handle
on its regulatory plate. Typically, some issues that don't seem
to be very important now will likely have an important
impact on wireless carriers and users in the near future. For
example:

   • Local number portability (LNP), which allows customers who change service providers to keep their telephone
numbers, seems simple enough, but its implementation
requires costly network upgrades for carriers and complex
call routing systems. In fact, according to a Boston-based
telecommunications consulting firm called The Yankee
Group, over the next few years no other single issue will cost
wireless carriers more in terms of labor and dollars.
   Wireless number portability will affect most network and
operational elements of wireless carriers, including customer
acquisition and retention, billing systems, intercarrier agreements, and network infrastructure. By 2002, The Yankee
Group estimates that carriers may collectively spend up to $1
billion upgrading their networks for number portability.

Beyond that, it is likely the carriers will continue to incur significant ongoing costs.

The Telecom Act of 1996 required the FCC to set standards for LNP and it has—setting a June 30, 1999, deadline for wireless carriers to implement number portability throughout their networks. Meanwhile, technical standards must be set for wireless LNP, and the carriers were still working on those issues in mid-1998.

• Calling party pays is another hot issue in the cellular industry because it obviously limits the use of cellular phones. As its name implies, the person who places a call pays for that call. This system is used throughout much of the world for wireless calls, but not in the United States, where the opposite is true. In the United States, airtime is paid by the person receiving the call on a cellular phone.

In February 1998, the Cellular Telecommunications Industry Association (CTIA) asked the FCC to rule that states cannot have inconsistent rules regarding calling-party-pays systems. The CTIA, which represents U.S. wireless service providers, says that such systems operate across state lines and could not logically be subject to different rules in different states.

• Mergers and consolidations in the industry, especially the increasing number of them, are another area directly related to telecom deregulation that will have to be dealt with on an ongoing basis. Clearly, size has its advantages in offering myriad telecom services. Everyone knows this from the results of surveys that indicate that business users and consumers prefer "one-stop shopping" when it comes to meeting their telecom requirements. Most survey respondents say they want voice, data, video, and even their entertainment on one bill. What concerns the carriers is that this attitude could change once their customers receive an "all-in-one" bill. The customers, they fear, may opt to drop one or more services or switch to a less costly carrier for certain services.

• Electronic commerce is a major issue unfolding. Congress continues to be under pressure to speed the growth of electronic commerce, and passage of the Telecom Act has not gotten Congress or the White House off the hook. The extraordinary pace of new technological developments and the rate of growth of telecom services will likely keep lawmakers busy for the foreseeable future. For the industry, this means a moratorium on Internet taxes and tariffs and a loosening of encryption-export controls—issues of some importance to the wireless community as it works to promote wireless data and mobile computing applications worldwide.

More Laws for More Issues?

If business, and even consumers, are going to improve their productivity and lifestyles, the industry argues, they will need fewer restrictions on how they can use these emerging technologies and lower cost services.

Selling Congress may be more difficult than selling the end user. In fact, the Telecom Act has not lived up to its promise of lower phone rates and better service. The big news has been the development of a number of huge mergers and acquisitions by some of the biggest telecom companies in the business.

When SBC Communications made a $56.18 billion bid to buy Ameritech Corp., FCC Chairman William Kennard issued a statement: "The bottom-line question," he said, "is, 'Is this merger going to create competition, or will it be a nonaggression pact?' The Telecom Act was all about opening markets for competition. SBC and Ameritech must show us that this merger will serve the public interest and enhance competition." While the results aren't in yet and may not be for years, many members of Congress and even some special interest telecom industry groups have begun to question the effectiveness of the new Telecom Act.

Congress does not seem to be in the mood to rewrite the legislation it fought so hard to get passed. But it may have to do some tweaking of the law if there is ever to be any hope of better meeting the realities of the market.

CHAPTER 7

What's Next?—Smaller, Lighter, Multifunctional

Y ou may have seen the ads. One of them shows a pair of multimedia glasses powered by solar panels along the stems, an infrared network link, a retinal display (for videoconferencing, multimedia, and web browsing), and a bone conduction transducer (for audio output)—all powered by very tiny batteries and an even smaller antenna.

Another ad depicts a highly stylized wristwatch with a built-in phone/TV so you can see who you are talking to and a highly miniaturized built-in camera so that you can be seen at the other end of the conversation.

These are not real products, but they're not very far off. The development of multimedia concepts for wireless/portable communications is well under way, as are advances in the integration of voice, data, fax, the Internet, video, and other imaging capabilities—all in the same highly miniaturized device. A few innovative companies are already experimenting with transmitting voice, data, and video over the same network. Among the research and development efforts:

• Ericsson is developing a system for transmitting video images from portable camcorders via the digital GSM network, and it has delivered a next-generation wideband CDMA system for mobile communications to NTT DoCoMo, Japan's largest mobile network.

• Kyocera Corp., another Japanese company, is marketing a Personal HandyPhone (PHS) visual phone adapter, or Datascope, that turns the Japanese PHS phone handset into a television receiver merely by attaching the optional adapter to a card slot on a PHS data communications terminal and using the PHS network. This PHS is shown in Figure 7-1.

• Sony Electronics' Wireless Telecommunications Co. is working on replacing the tape drive in its camcorders with a radio transmitter to send digital still and video images to a home- or business-based DVD for storage and later viewing. Sony has already demonstrated a CDMA-based concept phone with the capability to deliver voice, Internet access, and local information (e.g., traffic reports, movie schedules, maps, and yellow pages), along with digital photography.

• Toshiba Corp., Toyota Motor Co., and Fujitsu Ltd. have tied in with Alpine Electronics, Denso, Kenwood, Nippon TV, and Tokyo FM Broadcasting to launch a digital broadcast satellite (DBS) system that would use Motion Picture Expert Group (MPEG) technical standards used for mobile applications to broadcast TV-quality video and audio directly to Japanese vehicles. The Japanese companies have formed a broadcast company and plan to begin multichannel operations by 2001.

Figure 7-1. Kyocera Corp.'s Personal HandyPhone.

• Motorola has developed a video-streaming product called TrueStream, which delivers high-quality audio and video over a range of transmission speeds, including limited bandwidth connections. A useful tool for business users, video streaming means that users can see and hear video clips as they are downloaded without having to wait for the entire file to arrive. Transmitting the same information wirelessly may be the next step.

• Loral Orion Network Systems, a subsidiary of Loral Space and Communications, has introduced an Internet access service that will allow companies to maintain their existing terrestrial Internet access while expanding their capability to easily download large data files from the World Wide Web (WWW). The service is expected to be especially useful for bandwidth-intensive applications such as multicasting, downloading video, and audio streaming.

Challenges in Creating Multimedia Applications

Bandwidth is a major technical issue, especially for multimedia applications. A study of broadband and multimedia development and applications by the Chicago-based International Engineering Consortium (IEC) says that "future success for telecommunications companies rests on multimedia services supported by broadband capabilities." And while the IEC believes that the economic promise of multimedia applications is high, so are the risks. As the IEC points out in its study, "The road ahead is unclear. With competition for end users becoming more intense, creating and maintaining a competitive advantage is crucial."

Current wireless data systems are limited to low-bandwidth services and are insufficient for mobile multimedia. Kyocera's Datascope, for example, transmits at the rate of two frames per second—the conversation from the phone is heard over attached earphones, which means that it is not quite ready for prime time. The Ericsson system, demonstrated publicly for the first time in the fall of 1997, also could produce only slow-scan video images.

Such slow progress in this area of development among individual organizations may explain why so many heavy-weight wireless communications companies have been pooling their resources—forming strategic alliances for developing multimedia applications. Examples of strategic alliances that are increasingly common include the following:

- Motorola plans to integrate Florida-based NetSpeak's Internet protocol telephony technology into its wireless infrastructure and subscriber devices to support real-time multimedia communications—including voice, fax, audio, video, and data—over a wireless network.

- Sun Microsystems and Harris Semiconductor are jointly developing a wireless local area network that is expected to transmit Internet video five times faster than conventional wireless LAN products.

- AT&T and Excite, Inc., a global media company that offers consumers a free online, advertiser-supported Internet service, are developing an Internet-based multimedia communications service and have established a "personal communications center" that will enable AT&T's Digital PCS subscribers to send short messages over the Internet to other Digital PCS subscribers.

- Fujitsu has proposed a mobile device with eighteen keys designed for one-handed typing. Fujitsu has developed a prototype for 3Com's PalmPilot for the U.S. market. The keypad uses fourteen keys for alphabet characters, twelve of which double up on characters. The four remaining keys are function keys.

The Future Is Now

In the not-too-distant future, mobile phones will unlock and start cars, open front doors, or start dinner as you near your house in the evening. AT&T has been testing a fixed wireless system in Chicago that eliminates the need to run wires from utility poles to buildings. Other examples:

• *Voice recognition.* Keyboards may no longer be necessary as voice-recognition technology will let people dial a phone number simply by saying aloud a code name or the actual name (prestored in the device's memory) of the person they want to call, for example, "Mom."

• *Wristwatch communicators.* Several wristwatch/pager models are already on the market, mainly in the United States and Japan. A good indication of the direction some products will take has been demonstrated by AT&T Bell Laboratories (now a part of Lucent Technologies) with a highly miniaturized Dick Tracy concept phone with two wristbands. When the phone/watch receives a call, an electronic bell sounds or the user feels a vibration on the wrist, very much like most pagers work today. The user then presses down on the top of the watch to release an outer band that contains the speaker. This automatically takes the phone "off the hook." The speaker band pivots into the palm of the hand so that the user can cup the hand over the ear and speak into the wrist unit for a private conversation. When the call is completed, the speaker band can be snapped back into place over the first wristband.

Nippon Telegraph & Telephone Corp. (NTT), Japan's largest telecommunications company, loaned a newly developed wristwatch phone to several members of the organizing committee at the 1998 Winter Olympics in Nagano, Japan. They liked it so much that NTT said it planned to introduce a version of the watch/phone (based on Japan's digital mobile Personal HandyPhone System) by the end of the year.

- *Wireless teleconferencing/broadcasting.* Digital Equipment Corp.'s Mobile Computing and Communications Appliance (MOCCA) organization, meanwhile, has demonstrated a concept phone/computer with wireless teleconferencing that operates entirely on spoken commands. No keyboard is necessary. One of the key features of the device—which is four-inches square and can be worn around the neck or propped up on a desk—is an Internet web browser.

  A research project called Mobile Television and Innovative Receivers (MOTIVATE) and led by Deutsche Telekom has been organized as part of a European initiative headed by Esprit, the European Union's collaborative technology research program. MOTIVATE hopes to develop multimedia systems for vehicles and laptop computer users no later than 2002. Initially, the work will focus on the mobile reception of digital terrestrial and digital video broadcasting signals at fairly high data rates.

- *Virtual displays.* One question that product designers started asking themselves a few years ago is how they could reconcile consumers' and business users' demands for smaller and lighter wireless phones with the need to display more information. One likely possibility is an emerging technology development called "virtual displays," which uses camera-type optically aided viewfinders built into the phone to make the display screen seem larger than it actually is. For example, an architect using a cellular phone with a virtual display could talk to a colleague in his office about changes in a blueprint while standing miles away at a construction site for a new office building. The architect could discuss the changes while viewing the entire drawing through the eyepiece that is built into the phone.

  The same type of displays can also be used in pagers. You would simply hold the pager up to your eye to view a complete page of text, graphical information (including charts, tables or graphs), or even photographs.

Another display development recently demonstrated by Sharp Corp. and Semiconductor Energy Laboratories of Japan is a technology called continuous grain silicon (CGS), which could result in paper-thin, credit-card-size communications devices and computers. Significantly, Sharp says that both displays and chips could be produced using the CGS technology.

New Worldwide Standards Ahead

Where do we go from here? Frankly, it is hard to say. We have become so used to truly amazing advancements in technology that we have come to accept the rapid obsolescence of many of the products we buy. This is particularly true with communications devices and computers.

With European telecom regulators trying to stay ahead of U.S. and Asian competitors in adopting new wireless technologies, and with the rapid growth of European wireless markets, the EU estimates the European wireless market for cellular services will top $100 billion by 2005 and grow to 300 million users by 2015.

3G Mobile Telecommunications Standards

It is those kinds of numbers that have pushed the wireless community into developing so-called third-generation (3G) mobile telecommunications concepts and designs, which are moving closer to reality now that the European Telecommunications Standards Institute has reached a consensus agreement for a 3G mobile phone standard. The United States, Japan, and Korea, meanwhile, have submitted their own proposals to the International Telecommunications Union (ITU), the Geneva-based telecom regulatory arm of the United Nations, that promotes a somewhat different set of worldwide standards for future wireless networks. ETSI's influence, however, has made its proposed Universal Mobile Telecommunications System (UMTS) a leading candidate under the ITU's program to establish a new world 3G standard, known as International Mobile Telecommunications 2000, or IMT-2000.

IMT-2000

IMT-2000 is a highly ambitious program. It aims to integrate the various satellite, terrestrial, fixed, and mobile systems currently being deployed and developed under an umbrella standard that will pave the way toward true global service capabilities and interoperability soon after the year 2000.

The goal of 3G mobile systems is to provide users with worldwide coverage via handsets that can seamlessly roam between multiple networks—fixed and mobile, cordless and cellular—across regions that currently use different technologies. IMT-2000 will support a wide range of handsets, from simple messaging units through desktop multimedia terminals, and will be able to connect different radio transmission modules to the same core network equipment.

One of the most important 3G features will be high-speed wireless data, which will play a critical role in multimedia applications. Today's wireless data technology offers speeds in the 14 Kbps (kilobits per second) range. Initial research by Sprint PCS and others has demonstrated the potential for wireless data speeds as high as several megabits per second.

The present work schedule of the ITU calls for the "key choices" of radio transmission technologies associated with IMT-2000 to be made by March 1999, with appropriate ITU recommendations to be completed by the year 2000.

These dates could easily slip, however, if the U.S. Congress enters the debate over global wireless communications standards. Lawmakers, pressed by U.S. vendors and service providers, do not want American-born technologies, such as CDMA, locked out of 3G global wireless specifications. They also want to ensure that any wireless standard is compatible with current networks. Their fear is that the adoption of GSM, the European cellular standard, as the global 3G standard could fairly easily shut out CDMA from the 3G market.

Much is at stake in the 3G standards issue, with some of the largest manufacturers fighting for their technologies and with some high-level government lobbying, including significant involvement by the U.S. State Department. Accusations abound, led mainly by Qualcomm and Motorola in the United States and Ericsson in Europe. Japan has also made a strong bid to promote a single standard. By one knowledgeable account, ten technical standards have been proposed as the 3G wireless technology, with a few additional flavors thrown in by the Chinese, Koreans, and several others. In the United States, the Telecommunications Industry Association (TIA) has been charged with coordinating and presenting the U.S. proposals to the ITU for consideration.

3G Product Development

The convergence of communications and computers as well as the growing influence of the Internet and emerging multimedia devices are already challenging product developers to devise new strategic technology development programs.

The focus is on higher performance (meaning longer battery life), high data rate systems at low cost. In time, we will have access to downloadable software for multimode/multiband wireless phones with much greater functionality than is available today.

The potential for 3G applications may stretch the imagination for years to come, but the current list of possibilities is impressive enough. It includes:

- Interactive news delivery (e.g., voice, video, e-mail, graphics)

- Voice/high-quality audio

- Digital still photography

- Video

- Data transmission services

- Internet gaming

- Interactive audio

- File transfers from the Internet

- Voice/CD-quality music

- Multimedia e-mail (e.g., graphics, voice, video)

- Videoconferencing

- Web browsers

- Online services

- Time schedules

- Global positioning services/geographical information services

The key elements of most of these products and applications, however, will be a full range of Internet and wireless multimedia services.

A Long-Range Goal: Software-Definable Radios

As we have already pointed out at some length in earlier chapters, while the shift from desktop to portable computer and communications products has fueled many innovations in wireless technology, it also has led to the development of an increasing number of noncompatible communications services (e.g., CDMA, TDMA, and GSM, as well as at least three paging standards).

The impact of the wireless interoperability problem has extended to government agencies at all levels. For example, many agencies—the military services, federal, state, and local law enforcement agencies, fire departments, and other emergency services—occasionally need to communicate with each other but can't. The bombings of the Murrah Federal Building in Oklahoma City and New York City's World Trade Center are unfortunate examples of the problem. Since they all use different radio frequencies and normally operate only in those frequencies, FBI and Secret Service agents, local and regional police and fire officials, and other emergency services personnel could not communicate with each other at the scene of the bombings.

Multifunction Information Transfer System Forum

Sparked by these two incidents and the successful demonstration in 1995 of the U.S. Department of Defense's SPEAKeasy program, an international group of wireless communications companies and government agencies have formed an organization known as the Multimode Multifunction Information Transfer System (MMITS) Forum. The goal of this group is to develop and adopt an open technical architecture for a multiband, multimode, multifunctional, and multistandard wireless communications system. The architecture would provide both internal and external interface specifications that vendors can use as a guide for building hardware and software modules that service providers can use to link their networks to take full advantage of MMITS-developed technologies.

The forum is working to accelerate the development and deployment of software-definable radio systems to provide a system with multiple capabilities and flexibility for voice, data, messaging, multimedia, and future needs. Software-defined radios would use adaptable software and flexible hardware platforms to address the problems that arise from the constant evolution and technical innovation in the wireless industry. Another goal of the MMITS Forum is to reduce costs by increasing the demand for common modules such as radio frequency (RF) devices.

Business users will likely demand increased functionality and flexibility in their wireless services. Service providers cannot realistically replace their wireless networks every few years to keep up with the evolution of technology and standards. Technologies such as those being proposed by the MMITS Forum are likely to be an important strategy in the continuous evolution of the wireless network infrastructure.

Bluetooth Proposed— Will Anyone Bite?

Meanwhile, several industry leaders, including Ericsson, IBM, Intel, Nokia, and Toshiba, have come up with a program that would enable wireless service subscribers to connect a wide range of devices easily and quickly, expanding their communications capabilities for mobile computers, mobile phones, and other mobile devices, both in and out of the office.

Code-named Bluetooth, the program is being developed through the combined contributions of its member companies. Operating under the Bluetooth Special Interest Group (SIG) banner, the group hopes to develop the concept into a technical standard. The participating companies will each bring their own special expertise to the program.

Ericsson and Nokia, for example, will contribute their basic radio technology. Toshiba and IBM are developing a common specification for integrating Bluetooth technology into mobile devices. Intel is contributing its advanced chip and software expertise. Other companies are being invited to support the core technology on a royalty-free basis to ensure that Bluetooth can be implemented in many different devices.

How Bluetooth Works

Bluetooth uses a short-range radio link to exchange information between mobile phones, PCs, handheld computers, and other peripherals. The radio will operate on the globally available 2.4 GHz ISM (industrial/scientific/medical) "free band," allowing international travelers to use Bluetooth-enabled equipment worldwide.

Bluetooth is expected to eliminate the need for business travelers to purchase or carry numerous, often proprietary cables by allowing multiple devices to communicate with each other through a single port. Enabled devices will not need to remain within line of sight and can maintain an uninterrupted connection when in motion, or even when placed in a pocket or briefcase.

Bluetooth-enabled users will also be alerted to, and can respond to, incoming e-mail, even while their mobile PC remains in its carrying case. And they will be able to access the Internet via a completely wireless connection routed either through a mobile phone or a wired connection, such as the public switched telephone network (PSTN), integrated services digital network (ISDN), or a local area network (LAN). Business users will also have the capability to transmit pictures (e.g., a facility being considered for acquisition) to any location in the world by cordlessly connecting a camera to a mobile phone or any wire-bound connection—and add comments to their images using a mobile phone or mobile PC.

While Bluetooth is aimed at growing the market for personal mobile devices and increasing airtime usage, it may be just the concept needed for business people who actually want to or need to stay in touch with their colleagues—anywhere, anytime.

A Good Time to Be Wireless

This is a very good time for business users to take advantage of new wireless communications products and services. Market research and consulting firm Kenneth W. Taylor & Associates says we are now in the "massive cutthroat competition and new leadership" phase of the market, during which some of the early entrants in the wireless industry will take a beating from new entrants who have superior investment partners, technology-based innovations, and fast-reacting top management that can effectively cope with and survive what have become fairly routine price wars.

As competition from other wireless services heats up, cellular carriers may have to find new ways to attract business. The introduction of PCS throughout most of the country is an obvious example. Paging has begun to offer many of the same services available to cellular and PCS phone users, including two-way messaging, and it is much cheaper and works just fine for people who don't feel the need to actually talk to anyone, anywhere, anytime. Specialized Mobile Radio (SMR) operators still have a long way to go to catch up with most of the cellular systems in business today. However, by reinventing themselves—from a commercial dispatch service to a cellular-like digital network—the few big SMR operators are making a good run at providing a service targeted specifically at business users.

Wireless and the Economy

Clearly, the role of wireless will evolve over the next several years with many changes in the technology and regulatory environment. New applications will emerge. Rules and regulations will change. Competition at all levels will intensify. The rollout of PCS over the next several years could easily create up to six new choices for mobile users throughout the United States. All of these things will serve to redefine telecommunications services as we know them today.

Indeed, a new organization, the Global Mobile Commerce Forum, has been formed to promote the development across industries of electronic commerce services for mobile phones. The thinking of the forum is that mobile commerce will help to make both mobile telephony and electronic commerce more attractive to the mass market. Retailers and banks, for example, will use the mobile phone as a retail output in customers' pockets.

As cellular (particularly digital cellular) use continues to grow—and it continues to grow at the rate of 28,000 new subscribers a day in the United States—Internet activity will grow right along with it. A U.S. Commerce Department study, The Emerging Digital Economy, published early in 1998, found that traffic on the Internet has doubled every 100 days and that Internet commerce among business users will likely surpass $300 billion by the year 2002.

This is all part of the brave new world of wireless communications.

One thing is for certain: You will not be doing business in the future as you have in the past. Wireless communications devices will be every bit as ubiquitous tomorrow as personal computers are today. Wireless communications should be viewed in the same context—as a vital business tool that, when properly applied, can measurably enhance productivity.

Industry Sources

For additional information, you can contact the following wireless communications organizations. . . .

AIM USA
634 Alpha Dr.
Pittsburgh, PA 15238
(412) 963-9047
(A trade association dedicated to automatic identification manufacturers and applications, such as remote bar coding, ID tagging, and keyless entry.)

Alexander Research Co.
4854 E. Onyx Ave.
Scottsdale, AZ 85225
(602) 948-8225
Fax: (602) 948-1081
(A market consultant and research firm specializing in studies of how businesses use wireless communications technologies.)

American Mobile Telecommunications Association
1150 18th St. NW
Suite 250
Washington, DC 20036
(202) 331-7773
(A trade association for industrial and commercial communications interests.)

APCO International
2040 South Ridgewood Ave.
South Daytona, FL 32119
(800) 949-2726
(A trade association for law enforcement communications
officers.)

Association for Interactive Media
1019 19th St., NW
10th Floor
Washington, DC 20036
(202) 408-0008
Fax: (202) 408-0111
(An association formed to represent the interests of business
users of interactive video and audio media.)

CDMA Development Group
650 Town Center Dr.
Suite 820
Costa Mesa, CA 92626
(714) 545-9400
Fax: (714) 545-8600
(An organization dedicated to promoting the development
and use of code division multiple access technologies for
cellular, PCS, wireless local loop, and other CDMA-based
wireless applications.)

Cellular Telecommunications Industry Association
1250 Connecticut Ave., NW
Suite 200
Washington, DC 20036
(202) 785-0081
(The largest trade association of cellular and PCS carriers
and manufacturers.)

Center for the Study of Wireless Electromagnetic Compatibility
University of Oklahoma
202 W. Boyd
Suite 23
Norman, OK 73019
(405) 325-2429
(The primary site of basic research into electromagnetic
emissions from wireless communications sources in the
United States, including portable communications devices.)

Columbia Institute for Tele-Information
Columbia Business School
809 Uris Hall
New York, NY 10027
(212) 854-4222
Fax: (212) 932-7816
(A center for the study of advanced technologies and
applications and their use in business.)

Corporate Committee of Telecommunications Users (CCTU)
41 Fifth Ave.
New York, NY 10003
(212) 477-4377
Fax: (212) 477-6847
(Made up of small and large business users, as well as
universities and government agencies, the CCTU promotes
national telecommunications policies of interest to its
members.)

Electronic Industries Alliance
2500 Wilson Blvd.
Arlington, VA 22201
(703) 907-7500
Fax: (703) 907-7501
(A trade association representing the interests of a broad
spectrum of U.S. electronics manufacturers.)

Energy Telecommunications and Electrical Association
P.O. Box 639
Tomball, TX 77377
(A trade association serving oil and gas pipeline, gas
distribution, and electrical utility companies.)

Federal Communications Commission
1919 M. St. NW
Washington, DC 20055
(202) 632-7557
Fax: (202) 632-1587
(The primary communications regulatory agency of the U.S.
government.)

GSM MoU Association
Avoca Court
Temple Rd.
Blackrock Co
Dublin, Ireland
353 1 2695922
Fax: 353 1 2695958
(The lead organization for the worldwide development and
promotion of the Global System for Mobile Communications.)

Industry Canada
Spectrum Information & Telecommunications Division
235 Queen St.
Ottawa, Ontario K1A OH5, Canada
(613) 998-0368
Fax: (613) 952-1203
(Essentially, the FCC of Canada.)

International Association of Satellite Users and Suppliers
45681 Overbrook Court
Suite 107
Sterling, VA 20166
(703) 759-2094
(Represents satellite equipment suppliers, users, and relation
organizations.)

International Communications Studies Program
Center for Strategic & International Studies
1800 K St., NW
Washington, DC 20006
(202) 775-3263
Fax: (202) 775-0898

International Telecommunications Union
Place des Nations
CH-1211 Geneva 20, Switzerland
(41) 22 730 6161
Fax: (41) 22 730 6444
(The lead regulatory agency for international telecommuni-
cations, the ITU is primarily concerned with the allocation of
spectrum for global communications services.)

ITS America
400 Virginia Ave., SW
Suite 800
Washington, DC 20024
(202) 484-IVHS
Fax: (202) 484-3483
(The U.S. trade association of intelligent transportation systems developers and manufacturers of in-vehicle navigation systems, automotive radar, and related products and technologies.)

Multimedia Telecommunications Association (formerly the North American Telecommunications Association)
2500 Wilson Blvd.
Suite 300
Arlington, VA 22201
(800) 799-MMTA
Fax: (703) 907-7478
(A trade association of business communications equipment manufacturers.)

National Cable Television Association
1724 Massachusetts Ave., NW
Washington, DC 20036
(202) 755-3550
(The lead trade association of the cable television industry.)

National Care Information and Management Systems
Society (HIMSS)
230 E. Ohio St.
Suite 500
Chicago, IL 60611
(312) 664-4467
Fax: (312) 664-6143
(HIMSS members analyze and design hospital telecommunications information systems.)

National Retail Federation
325 7th St.
Suite 1000
Washington, DC 20004
(202) 783-7971
Fax: (703) 737-2849
(Represents state and federal retail associations and their
employees in all phases of retailing, including telecommunications and data processing activities.)

National Telecommunications & Information Administration
U.S. Department of Commerce
Washington, DC 20230
(202) 377-1866
Fax: (202) 501-6198
(The U.S. agency that advises the Administration of U.S. telecom policy, and which represents U.S. government agencies
in telecommunications issues, particularly spectrum allocation.)

Personal Communications Industry Association
1501 Duke St.
Alexandria, VA 22314
(703) 836-3528
Fax: (703) 836-3528
(Originally the trade association for the U.S. paging industry,
PCIA membership has branched out to include cellular and
other wireless communications interests.)

Rural Cellular Association
2711 LBJ Freeway
Suite 560
Dallas, TX 75234
(800) 722-1872
Fax: (214) 243-6139
(A trade association representing rural cellular interests.)

Satellite Broadcasting and Communications Association
225 Reinekers Lane
Suite 600
Alexandria, VA 22314
(703) 549-6990
(A trade association for the satellite communications
industry.)

Scientific Advisory Group on Cellular Telephone Research
1711 N St., NW
Suite 200
Washington, DC 20036
(202) 833-2800
(An organization formed by the cellular industry to study
all aspects of the impact of wireless communications on
human health.)

Telecommunications Industry Association
2500 Wilson Blvd.
Suite 300
Arlington, VA 22201
(703) 907-7700
Fax: (703) 907-7727
(An association of communications equipment manufacturers, the TIA is involved in the development and promotion of
industry standards. The TIA is closely associated with the
Electronic Industries Alliance, formerly the Electronic Industries Association.)

U.S. GPS Industry Council
1100 Connecticut Ave., NW
Suite 535
Washington, DC 20036
(202) 296-1653
(The trade association of the global positioning system
industry.)

U.S. Telephone Association
1401 H St., NW
Suite 600
Washington, DC 20005
(202) 326-7300
(The trade association of wireline telephone carriers.)

UTC—The Telecommunications Association
1140 Connecticut Ave., NW
Suite 1140
Washington, DC 20036
(202) 872-0030
(A trade association representing the telecommunications
interests of the U.S. utilities industry.)

Utilities Telecommunications Council
1140 Connecticut Ave, NW
Suite 1140
Washington, DC 20036
(202) 872-0030
Fax: (202) 872-1331
(An organization of gas, electric, water, and steam utilities
involved in developing telecommunications operations of
utilities.)

Wireless Cable Association International
1140 Connecticut Ave., NW
Suite 810
Washington, DC 20036
(202) 452-7823
(The trade associaton of the wireless cable industry.)

Wireless Data Forum
1250 Connecticut Ave., NW
Suite 200
Washington, DC 20036
(202) 736-3663
(An industry association that promotes the use and advance-
ment of cellular digital packet data technologies as an international standard.)

Wireless LAN Interoperability Forum
1111 W. El Camino Real
Suite 109-171
Sunnyvale, CA 94087
(415) 960-1630
(A group of industry companies organized to promote the
interoperability and compatibility of wireless local area
network equipment.)

Glossary of Terms and Acronyms

A

"A" Carrier   The nonwireline cellular company that operates in radio frequencies from 824–849 MHz. (See "B" Carrier.)

Access fee   A special fee that local phone companies charge their customers for the right to connect with the local telephone network. The fee is paid by cellular subscribers.

Access point (AP)   A centrally located transceiver (transmitter/receiver) that provides wireless terminal nodes access to the local wireless network.

Advanced intelligent networks (AIN)   Systems that allow a wireless user to make and receive phone calls while roaming in areas outside the user's home network. These networks, which rely on computers and sophisticated switching techniques, also provide many personal communications service features.

Advanced Mobile Phone System (AMPS)   The U.S. technical standard for analog cellular telephones.

Airtime   The actual time spent talking on a cellular or PCS phone, which is usually billed to the subscriber on a per-minute basis.

Allocation   The designation of a band of frequencies for a specific radio service or services. The Federal Communications Commission (FCC) and the National Telecommunications and Information Administration (NTIA) are responsible for frequency allocations in the United States.

Alphanumeric   A message, usually from an alphanumeric pager, that displays both letters ("alpha") and numbers ("numeric").

Analog   The traditional method of transmitting voice signals where the radio wave is based on electrical impulses that occur when speaking into the phone.

ANSI   American National Standards Institute

ATM   Asynchronous transfer mode

B

"B" Carrier   The wireline cellular carrier, usually the local telephone company, that operates in the frequencies 869–894 MHz.

Bandwidth   The total range of frequencies required to transmit a radio signal. The bandwidth of a radio signal is determined by the amount of information in the signal being sent.

Base station   The fixed transmitter/receiver station used by cellular and other wireless carriers to send, receive, and transport signals and to establish a communications link to the public switched telephone network (PSTN).

Basic Trading Area (BTA)   A service area defined by Rand McNally and adopted by the Federal Communications Commission (FCC) to promote the rapid deployment and ubiquitous coverage of personal communications services

(PCS) and providers. There are 493 BTAs in the United States.

Byte rate   The speed of a digital transmission, measured in bits per second

Broadband   A communications channel with a bandwidth greater than 674 Kbps (kilobits per second) that can provide high-speed data communications via standard telephone circuits. Also referred to as *wideband.*

Broadband PCS   Frequencies auctioned by the Federal Communications Commission (FCC) in the radio band of 1,850–1,990 MHz

Bundling   Usually, the sale of a cellular phone at a reduced cost only if you get a cellular service at the same time. Bundling is not legal in all states.

C

Call forwarding   A feature that lets you forward your phone calls to the telephone number of your choice

Call log   A feature that allows you to see the last ten incoming and outgoing calls on your phone display screen and to speed-dial any number in the log

Call waiting   A feature that allows you to answer an incoming call while on another call

CAP   Competitive access provider

Carrier   A cellular or personal communications service (PCS) service provider

Cell   The geographic area served by a single low-power transmitter/receiver. A cellular system's service area is divided into multiple/overlapping cells.

Cellular digital packet data (CDPD)   Introduced in 1992 and now available in most geographic areas of the United

States, CDPD uses the idle time in the analog cellular telephone system to transmit data in "packets" at rates up to
19.2 Kbps. Compare *circuit-switched data.*

Centrex   The switching system of a local telephone operator

Channel   A single path of the spectrum taken up by a radio
signal. A channel is usually measured in kilohertz (kHz).
Most analog cellular phones use 30 kHz channels.

Chip   An electronic component part. It can be either passive
or active, discrete, or an integrated circuit (IC).

Circuit-switched data   Unlike CDPD, transmitting circuit-
switched data is made possible by keeping a circuit open
between uses for the duration of a connection. Compare
CDPD.

Code division multiple access (CDMA)   CDMA is a digital
technology that uses a low-power signal that is spread
across a wide bandwidth. With CDMA, a phone call is
assigned a code rather than a certain frequency. Using the
identifying code and a low-power signal, a large number
of callers can use the same group of channels. CDMA proponents estimate its capacity is as much as ten times
higher than analog transmissions.

COMSAT   Communications Satellite Corporation, a private
company formed by Congress in 1992 that provides satellite communications capacity to carriers such as AT&T
and MCI, and other major satellite users.

Commercial Mobile Radio Service (CMRS)   The regulatory
classification that the FCC uses to govern all commercial
wireless service providers, including cellular, personal
communications service (PCS), and Enhanced Specialized
Mobile Radio (ESMR).

Common carrier   A government-regulated company that
provides telecommunications services in a specified

region. Common carriers include the telephone companies as well as the owners of communications satellites.

CT-1 Industry shorthand for cordless telephone/first-generation phones

CT-2 Cordless telephone/second-generation, a digital cordless telephone standard. These phones can be used for residential or business service.

CTIA Cellular Telecommunications Industry Association

D

dB Decibel, a unit for measuring the relative strength of a signal parameter, such as power or voltage

DCS-1800 Digital communications service at 1,800 MHz. DCS-1800 is a variant of the digital Global System for Mobile Communications (GSM) standard and is used outside the United States.

Digital A method of transmitting voice or data using the computer's binary code of 0s and 1s. For wireless communications purposes, digital transmission offers a clearer signal than analog technology. Cellular systems providing digital transmission are currently in operation throughout most developed regions of the world and will eventually become available in less developed areas.

Digital Advanced Mobile Phone System (D-AMPS) Digitally enhanced AMPS

Digital European Cordless Telecommunications (DECT) A digital cordless telecommunications system intended initially for wireless private branch exchange (WPBX) applications. However, this service may be used in the consumer market. DECT supports both voice and data communications.

Downlink   The transmission of radio frequency (RF) signals from an orbiting satellite to a ground communications station

Dual-band   A telephone handset that operates on 800 MHz cellular and 1,900 MHz PCS frequencies

Dual-mode phone   A phone that operates on both analog and digital networks

E

Earth (or ground) station   The electronic ground equipment that processes the radio frequency (RF) signals received from or sent to an orbiting satellite

EIA   Electronic Industries Association

EMI   Electromagnetic interference. Usually, unintentional interfering signals generated within or external to electronic equipment. Typical sources of interference could be power lines or electronic noise from certain types of power supplies and/or spurious radiation from other electronic sources.

Enhanced Specialized Mobile Radio (ESMR)   The next generation of SMR, a radio service traditionally used in the United States for dispatch radio services, such as taxi and fleet delivery services.

ESN   Electronic serial number. Every wireless phone must have a unique ESN and no two phones may have or emit the same ESN.

European Telecommunications Standards Institute (ETSI)   One of the European organizations responsible for establishing common, industrywide technical standards for telecommunications services in Europe

Exchange   A switching center, or the area where the common carrier furnishes service at the exchange rate and under the regulations applicable in that area, as prescribed in the carrier's filed tariffs

F

Federal Communications Commission (FCC)   The government agency primarily responsible for the allocation of radio spectrum for communication services in the United States

FLEX   A Motorola-developed messaging protocol that gives service providers more capacity and faster transmission times on their networks. Extensions of FLEX include InFLEX and ReFLEX.

Footprint   The area of the earth's surface covered by a satellite signal

FPLMTS   Future Public Land Mobile Telephone System. See also IMT-2000.

Frequency division multiple access (FDMA)   A multiplexing method whereby each channel is assigned a specific frequency for its use

Frequency reuse   Because of their low power, radio frequencies assigned to one channel in a cellular system are limited to a single cell. However, service providers are free to reuse the frequencies in other cells in the system without causing interference.

Full-duplex transmission   A communications circuit that can simultaneously transmit and receive information

G

Gateway   Essentially, an earth (or ground) station that links a satellite or satellites with the terrestrial public telephone network. Also, a part of a local area network (LAN) that allows it to interface with different transmission networks, particularly the public switched telephone network (PSTN).

Geostationary satellite (GEO)   A satellite whose speed is synchronized with the earth's rotation so that the satellite is always in the same location over the earth. Most geostationary satellites orbit the earth at an altitude of approximately 22,300 miles.

Gigahertz (GHz)   A frequency equal to one billion Hertz, or cycles per second

Global Positioning System (GPS)   A network of twenty-four satellites developed and operated by the U.S. Department of Defense that provides precise location determination anywhere in the world via special receivers. These receivers, which are being used increasingly for commercial (aviation, surveying) and recreational (boating and hiking) applications, can provide an accuracy of less than 100 meters.

Global System for Mobile Communications (GSM)   Originally called the Groupe Speciale Mobile, GSM is the European digital cellular standard. It has also been adopted by several service providers in other regions of the world, including the United States, for both cellular and PCS services.

H

Half-duplex transmission   A communications circuit that can transmit or receive information, but not simultaneously

Handoff   Cellular systems are designed so that a phone call can continue while driving from one area of cellular transmission coverage, or "cell," to another cell. The transfer of the call from cell to cell, which is normally transparent to the caller, is called the handoff.

Hertz   The unit of measuring frequency signals. One Hertz is equal to one cycle per second.

Handheld Device Markup Language (HDML)   A computer language that enables Internet access from wireless devices. These include handheld personal computers (HPCs) and smart phones.

High tier   A wireless system that may serve high-speed vehicular traffic and may have high-power levels

Home location register (HLR)   A network element containing a database of customer information that makes subscriber information available to a mobile telephone switching office

I

IAPP   Interaccess point protocol. Proposed protocol that would facilitate roaming across access points.

IEEE 802.11   The designation given the globally accepted technical standard for a wireless local area network (WLAN). IEEE 802.11 refers to the Institute of Electrical and Electronics Engineers subcommittee that developed and approved the operating characteristics and protocols of the WLAN standard.

IMT-2000   International Mobile Telecommunications (year) 2000

Industrial/scientific/medical (ISM)   This is the unlicensed radio band in North American and some European countries. Referred to in FCC documents as Part 15.247, it defines the parameters for use of the ISM bands in the United States, including power output and noninterference measurements. Commonly used ISM bands include 902–928 MHz, 2400–2483 MHz, and 5725–5850 MHz.

Infrastructure equipment   The fixed transmitting and receiving equipment in a communications system. It usually consists of a base station, base station controllers, antennas, switches, computers, and other equipment that makes up the backbone of the system that sends and receives signals from mobile or handheld subscriber equipment and/or the public switched telephone network (PSTN).

INMARSAT   The International Maritime Satellite Organization, which operates a network of international mobile communications satellites for maritime, aeronautical, and land mobile users

Integrated services digital network (ISDN)   A switched network providing end-to-end digital connectivity for simultaneous transmission of voice and data over multiplexed communications channels

INTELSAT   The International Telecommunications Satellite Organization. Formed in 1964, INTELSAT operates a global satellite system.

Interface   The point of interconnection between a wireless carrier's and a local exchange carrier's network facilities

Interference   Radiated energy that interferes with the reception of radio signals

International Mobile Telecommunications 2000 (IMT-2000)   An initiative undertaken by the International Telecommunications Union (ITU) to establish a global standard for third-generation wireless multimedia communications. Formerly referred to as Future Public Land Mobile Telephone System (FPLMTS).

Interoperability   The ability to migrate communications transmissions among a variety of local, regional, and national networks. Switching between the different networks would be transparent to the user.

IrDA   Infrared Data Association

IS   Interim standard

ISA   Industry Standard Architecture. An expansion bus that accepts plug-in boards on PCs. Wireless LAN adapters may be implemented as ISA cards. Compare *PCI*.

IS-41   Interim standard-41. The protocol for roaming within the United States, as designated by the Telecommunications Industry Association (TIA), which serves as the telecommunications arm of the American National Standards Institute (ANSI).

IS-54   Interim standard-54. The North American dual-mode (analog and digital) cellular standard. In the analog mode, IS-54 conforms to the AMPS standard.

IS-95   Interim standard-95. The code division multiple access (CDMA) standard for U.S. digital cellular service.

IS-136   Interim standard-136. The time division multiple access (TDMA) standard.

IS-661   Interim standard-661. A technology developed by Omnipoint, a PCS licensee in the New York Major Trading Area (MTA), which combines elements of code division multiple access (CDMA) and time division multiple access (TDMA) technologies.

ISO   International Standards Organization

ITS   Intelligent transportation systems

ITU   International Telecommunications Union

J

Japan Digital Cellular (JDC)   A digital cellular standard developed by Nippon Telegraph & Telephone (NTT) of Japan, operating in Japan at 800 and 1,500 MHz

JTAC   Japan's variant of the Total Access Communications System (TACS) analog standard. JTAC was developed for Japan by Motorola.

L

LATA   Local access transport area

LCD   Liquid crystal display

LEC   Local exchange carrier

LEO   Low-earth orbit satellite

Li-ion   Lithium ion (battery)

LNP   Local number portability, or customers' ability to keep their telephone numbers even if they change service providers

Local exchange area   The part of the national telephone network controlled by the local telephone operating company. Local exchange areas are generally regulated by the state public utility commission.

Local exchange carrier (LEC)   A wireline phone company serving a local area

Local multipoint distribution service system (LMDS) A new "wireless cable" service operating in the 28 GHz band. LMDS uses low-power transmitters, configured in a cellular-like arrangement, to transmit video and other services to special receivers in homes and businesses.

Local loop   The connection between a subscriber's telephone and the local telephone company's central office

Low tier   A wireless system intended for pedestrian or low-speed traffic and using low-power levels

M

Major Trading Area (MTA)   An area defined by Rand McNally and adopted by the FCC to designate personal communications service (PCS) coverage. There are fifty-one MTAs in the United States.

Megahertz (MHz)   A measurement of frequency equal to one million cycles per second. One cycle per second is the same as one Hertz.

MEO   Mid-earth orbit satellite

Microcells   Cell sites with small coverage areas. Antenna heights are generally low (forty feet or less).

Microwave   Radio frequency (RF) signals between 890 MHz and 20 GHz. Point-to-point microwave transmission is commonly used as a substitute for copper or fiber cable.

MIPS   Million instructions per second

Mobile telephone switching office (MTSO)   The central computer that connects a cellular phone call to the public telephone network. The MTSO controls the entire system's operations, including monitoring, billing, and handoffs.

Modem   A device that changes analog signals to digital signals for transmission over analog circuits, or changes digital signals to analog signals for transmission over analog circuits. Technically, modems modulate discontinuous digital signals into continuous analog waves for transmission over analog circuits and then demodulate the waves into digital bit streams at the receiving end of the circuit.

MPEG   Motion Picture Expert Group. Also refers to the actual technical standard for compressing full-motion video (as developed by MPEG).

MSS   Mobile satellite service

Multipoint   Three or more stations connected to the same facility

N

Narrowband PCS (N-PCS)   The designation given by the Federal Communications Commission (FCC) to nationwide frequency bands auctioned to operators of personal communications services (PCS). N-PCS operators are generally paging companies that wish to increase their capacity and add two-way capability. N-PCS frequency bands are not appropriate for cellular phone services.

National Telecommunications and Information Administration (NTIA)   An agency of the U.S. Commerce Department, the NTIA serves as the President's adviser on telecommunications policy and is responsible for administering all federal government use of the radio spectrum, including military communications.

NC   Network computer

NiCd   Nickel cadmium (battery)

NII   National Information Infrastructure

NiMH   Nickel metal hydride (battery)

NMT   Nordic Mobile Telephone

Node   A terminal point for two or more communications links. Nodes can serve as the control location for switching data in a network or networks.

Nonwireline cellular company   The FCC has licensed two cellular systems in each market (region)—one license for the local telephone company and the second (the "A" carrier) for other ("B" carrier) applicants. The distinction between A and B was meaningful only during the FCC's licensing process. Since these systems have been constructed, many of them have been sold. As a result, in some markets today, both the A and B systems are owned by telephone companies.

P

PABX   Public automatic branch exchange

PBX   Private branch exchange

Packet switching   A data transmission technique in which data messages are divided into blocks, or packets, of standard length. Each packet has address and control information coded into it.

PACS   Personal Advanced Communications Systems

pACT   Developed primarily by Bellcore, personal Air Communications Technology is a two-way paging and messaging protocol and system specification for Narrowband PCS.

PC   Personal computer

PCI   Peripheral Component Interconnect. A local bus for PCs for exchanging data between the computer's CPU and peripheral devices. The PCI bus coexists with the ISA bus. Wireless LAN adapters may be implemented as ISA cards.

PCIA   Personal Communications Industry Association

PCMCIA   Portable Computer Memory Card International Association. The abbreviation also refers to the credit-card-size cards used for attaching a modem, network adapter, or hard disk to a portable computer. A PCMCIA card is also called a PC card.

PCN   Personal communications network

PCS-1990   The U.S. variant of the European Global System for Mobile Communications (GSM) digital cellular standard that operates at 1,900 MHz.

PDC   Personal digital cellular (Japan)

Personal digital assistant (PDA)   A portable computing device capable of transmitting data. Most PDAs can be used for paging, data messaging, electronic mail, receiving stock quotations, and facsimile. They also function as personal computers and electronic organizers.

Personal HandyPhone System (PHS)   Japan's designation for its digital cordless telephony standard

PIN   Personal identification number

POP   A cellular industry term for population. If the coverage area of a cellular service provider includes a population base of one million people, it is said to have one million POPs. The financial community uses the number of potential users to help them measure the value of a cellular carrier.

Private branch exchange (PBX)   A telephone switching system designed to control and route calls in large multiphone environments, such as offices. Most PBXs are designed to handle custom features for users' specific telecom requirements.

Public switched telephone network (PSTN)   The domestic telecommunications network commonly accessed by most telephone systems, PBX trunks, and data networks.

R

Radio frequency (RF)   In terms of cellular applications, RF is the part of the electromagnetic spectrum between the audio and high-range frequencies (between 500 kHz and 300 GHz). Cellular transmission frequencies are found in two locations in the spectrum—at 824-849 MHz and 869-894 MHz.

RBOC   Regional Bell operating company

Reseller   Cellular service providers that purchase blocks of airtime at wholesale prices from cellular carriers and then resell the service at retail prices

RFID   Radio frequency identification

RISC   Reduced instruction set computer

Roaming   Using a cellular phone outside your usual service, for example, in a city other than the one in which you live

Rural service area (RSA)   One of the 428 Federal Communications Commission (FCC)-designated rural markets across the United States

S

Smart card   Credit card with built-in microprocessor and memory used for storing information, often for identification and financial transactions.

Speed dial   A cellular or PCS phone feature with memory locations activated by one or more keys to dial previously stored telephone numbers from memory

SONET   Synchronous optical network

Spectrum   The complete range of electromagnetic waves, which can be transmitted by such devices as cellular phones. Electromagnetic waves vary in length and therefore have different characteristics. Longer waves in the low-frequency range can be used for communications, whereas shorter waves of high frequency show up as light.

Spread spectrum   Originally developed by the U.S. military because it offers secure communications, spread spectrum radio transmissions essentially "spread" a radio signal over a very wide frequency band in order to make it difficult to intercept and difficult to jam. Spread spectrum techniques are becoming widely used in a number of commercial communications applications, mainly wireless local area networks (Walls).

Signaling System Number 7 (SS7)   An out-of-band digital switching network used internationally, mainly by local exchange carriers, to provide basic routing information, call setup, and other cell termination functions. In SS7, signaling is removed from the voice channel itself and put on a separate data network. In a wireless network, SS7 also provides subscriber identity and service usage through the validation using intelligent network elements, such as the Home Location Register (HLR) or Authentication Center (AUC).

Standby time   The amount of time a fully charged wireless phone can be left on before its battery runs down

Subscriber identity module (SIM)   A smart card that is installed or inserted into a mobile phone containing subscriber-related data, usually for billing purposes. Users with their own SIM card can borrow a SIM-enabled cellular phone and be billed for the use of that phone.

T

T1   The transmission bit rate of 1.544 megabits per second (Mbps) or the equivalent of the ISDN Primary Rate Interface for the United States. The European T1 or E1 transmission rate is 2.048 Mbps.

TCP/IP   Transmission Control Protocol/Internet Protocol. The Internet protocol suite developed by the U.S. Defense Department. It governs the exchange of sequential data, while IP routes outgoing and incoming messages.

Telecommunications Industry Association (TIA)   The U.S.-based organization established to provide industrywide standards for telecommunications equipment used in North America

Terrestrial communications   Any communications system where all transmitters and receivers are on the ground

Text messaging   Service that allows you to receive up to twenty-three alphanumeric messages of up to 100 characters each. In some systems, visual alerts let you know when you have messages.

Thin client   Terminals that provide little or no local storage and rely on the network for applications, data, and functionality; they primarily function as access devices. Thin clients combine wireless connectivity with corporate networks to access the applications and resources of a remote server or computer mainframe.

Time division multiple access (TDMA) TDMA increases the channel capacity by dividing the signal into pieces and assigning each piece to a different time slot. Current technology divides the channel into three time slots, each lasting a fraction of a second. Thus, a single channel can be used to handle three simultaneous calls. Compare *CDMA.*

Total Access Communications System (TACS) An analog cellular system used mainly in Europe. It has also been implemented in some areas in Japan, Britain, China, and other regions of the world.

Transceiver A radio that both transmits and receives radio signals

Trunk A communications channel linking a central office with a PBX or other switching equipment

U

U-NII Unlicensed National Information Infrastructure. Specifically refers to the FCC's spectrum allocation for short-range, high-speed wireless digital communications. The U-NII spectrum would support the creation of wireless local area networks for community use, mainly schools and libraries.

Universal Mobile Telephone Telecommunications System-Service (UMTS) An evolving third-generation, cellular-based mobile communications service, promoted primarily by European telecom interests

Universal service Essentially means providing basic voice telephone service at affordable rates to everyone in the country

Uplink The transmission of a radio frequency (RF) signal from an earth (ground) station to a satellite

V

Vocoder   A voice compression technique that employs algorithms so that only some of the digitally encoded voice signal is actually transmitted

W

WACS   Wireless Access Communications System (a Bellcore-developed service)

WARC   World Administrative Radio Conference

WCS   Wireless Communication Service. WCS is defined by the FCC as radio communications that may provide fixed, mobile, radio location, or satellite communication services to businesses within their assigned spectrum block and geographical area. WCS operates in the 2.3 GHz band.

Wireless access point   An access point connects the WLAN to the wired network and is similar in function to a wired server

Wireless local area network (WLAN)   A network that allows the transfer of data and the ability to share resources, such as printers, without the need to physically connect each node, or computer, with wires

Wireless local loop (WLL)   A wireless connection between the telecommunications service subscriber and the public switched telephone network (PSTN). WLL systems can be used instead of copper wires to connect telephones and other communications devices in the public telephone network.

Wireless private branch exchange (WPBX)   A standard office telephone system, designed for use on private premises such as offices, that has more extensions than incoming lines

WRC   Wireless Radiocommunications Conference. WRC is an international forum for developing and regulating the use of radio frequencies and satellite orbits.

Index

Advanced Mobile Phone Service
    (AMPS), 4
AIM USA, 137
AirBrowse, 93
aircraft, 28–29
"air interface" standards, 6–7
AirMedia, 93
Alexander Research Co., 137
American Mobile Satellite Corp.
    (AMSC), 56–57
American Mobile Telecommuni-
    cations Association, 138
American Radio Telephone
    Service, Inc., 3
American Stock Exchange
    (AMEX), 50–51
Ameritech Corp., 118
AMEX (American Stock
    Exchange), 50–51
AMPS (Advanced Mobile Phone
    Service), 4
AMSC, *see* American Mobile
    Satellite Corp.
Amtech Corp., 21
analog technology, digital
    technology vs., 4–5
APCO International, 138
ARDIS, 57, 79, 95, 101
"A" side carrier, 3
Association for Interactive

Media, 138
asynchronous transfer mode
    (ATM), 54
AT&T, 2–3, 13, 85, 91, 103
AT&T Digital PCS, 8
AT&T Wireless Services, 8, 38, 43,
    45

Basic Trading Areas (BTAs), 10
Bell Atlantic Mobile, 84
Bell Canada, 37
BellSouth Cellular Corp., 84
BellSouth Mobility DCS, 30–31
BellSouth Wireless Data, 79, 95
Big LEO systems, 57–59, 61, 64,
    65, 69–70
bin Talal, Walid, 64
Bluetooth program, 132–133
Boeing, *x*
broadband PCS, 10
broadcasting, wireless, 126
"B" side carrier, 3
BTAs (Basic Trading Areas), 10

car rentals, 83
Casio, 92
CCTU (Corporate Committee of
    Telecommunications Users),
    139
CDMA, *see* code division

169

multiple access
CDMA Development Group, 138
CDPD, *see* cellular digital packet
    data
Celestri System, 67
cells, 3–4
cellular digital packet data
    (CDPD), 79, 83–85, 93–95
cellular phones, 22–31
    on airplanes, 28–29
    fraud involving, 27–28
    health risks of, 24–27
    service pricing for, 22–24
    and smart cards, 30–31
cellular technology
    functioning of, 3–4
    history of, 2–3
    PCS vs., 7–9
Cellular Telecommunications
    Industry Association (CTIA),
    9, 117, 139
Center for the Study of
    Wireless Electromagnetic
    Compatibility, 139
CGS (continuous grain silicon),
    127
Clinton administration, 27, 75
code division multiple access
    (CDMA), 6–7, 37, 40, 115, 128
Coghill, Roger, 24
Columbia Institute for Tele-
    Information, 139
Communications Assistance for
    Law Enforcement Act, 27–28
computing, mobile, 89–101
    end-user devices in, 90–96
    and Internet-intranet access,
    90
    network, 98–101
    thin, 96–98
COMSAT, 59–61, 71–72
continuous grain silicon (CGS),
    127
cordless phones, 17–18
Corporate Committee of

Telecommunications Users
    (CCTU), 139
costs
    of digital wireless service, 5–6
    of satellite communications,
    61–63
    of WLANs, 48
    of WWL systems, 39–40
credit cards, 83–84
CruiseConnect, 100
CT-2, 8
CTIA, *see* Cellular Telecommuni-
    cations Industry Association

data, wireless, 77–88
    business applications for, 81–84
    and Internet, 87–88
    service availability for, 79–80
    slow user acceptance of, 77–78
    trends in, 84–88
DataTAC, 80
DBS (direct broadcast satellite),
    69
DC23 Mobile Office, 85
DECT (Digital European Cordless
    Telecommunications), 23
Delta Airlines, 73
digital camera, *x*
Digital Equipment Corp., 126
Digital European Cordless
    Telecommunications
    (DECT), 23
digital technology
    advantages of, 5–7
    analog technology vs., 4–5
digital TV, 122
direct broadcast satellite (DBS),
    69

ECCO, 58, 69
EC (European Community), 26
electromagnetic fields, 24
electronic commerce, 117–118
Electronic Industries Alliance,
    140

electronic serial number (ESN),
    28
Ellipso, 58, 69
e-mail, 16, 98–99
Energy Telecommunications and
    Electrical Association, 140
Enhanced SMR (ESMR), 11–14
Ericsson, 45, 80, 85, 121–122, 124
ESLPN Enterprises, 16
ESMR, *see* Enhanced SMR
ESN (electronic serial number),
    28
ESPNet, 16
Ethernet, 97
ETSI, *see* European Telecommu-
    nications Standards Institute
European Community (EC), 26
European Telecommunications
    Standards Institute (ETSI),
    114–115, 127
European Union, 111
EurtelTRACS, 72
EZ-Pass system, 21

FAA, *see* Federal Aviation
    Administration
FCC, *see* Federal Communica-
    tions Commission
FDA, *see* U.S. Food and Drug
    Administration
Federal Aviation Administration
    (FAA), 29, 75
Federal Communications
    Commission (FCC), 56,
    115–118, 140
  and Bell companies, 104–105
  and commercial radio, 2, 3
  and COMSAT, 60, 61
  and PCS, 9–11, 17
  and Project Angel, 38
  and RF safety, 25–26
  and Telecom Act, 107–110
  Wireless Communications
    Service of, 29–30
Federal Express, 82, 101

FoneBook PLUS, 91
Food and Drug Administration
    (FDA), 95
France, 37
fraud, 27
Fujitsu, 124
future trend(s), 121–135
  Bluetooth program as, 132–133
  in multimedia, 123–124
  in regulation, 116–119
  in satellite communications,
    54–61
  software-definable radios as,
    130–132
  in standards, 127–130
  teleconferencing/broadcasting
    as, 126
  virtual displays as, 126–127
  voice recognition as, 125
  in wireless data, 84–88
  in WLANs, 50
  in WLL, 37–38
  wristwatch communicators as,
    125

Gates, Bill, 58, 64
Georgia Hospital Health Services,
    44
geostationary orbit (GEO) sys-
    tems, 20, 54, 56, 57, 67, 69, 80
Geotek, 14
Global Positioning System (GPS),
    73–75
Globalstar, 58, 61–62, 67–69
Global System for Mobile
    Communications (GSM), 7,
    79, 94
GLONASS, 76
Goldman, Sachs & Co., 96
GPS, *see* Global Positioning
    System
GSM, *see* Global System for
    Mobile Communications
GSM MoU Association, 140
GTE, 92–93

Halliburton Brown & Root (HBR), 51
handheld personal computers (HPCs), 16, 89, 91–96
Harris Semiconductor, 124
HBR, *see* Halliburton Brown & Root
healthcare, 100–101
health risks, 24–27
hearing aids, 25
Hertz Rent-A-Car, 73–74, 83
Hewlett-Packard, 15
high-speed wireless services, 109–110
HIMSS (National Care Information and Management Systems Society), 142
history of cellular communications, 2–3
Hoechst Marion Roussel (HMR), 95
HPCs, *see* handheld personal computers
Hughes Network Systems, 45

IAPP (interaccess point protocol), 50
IBM Corp., *ix*
ICO Global Communications, 58, 66–67
iDEN (Integrated Digital Enhanced Network), 80
IEEE, *see* Institute of Electrical and Electronics Engineers
Illinois Power, 51
IMT-2000, *see* International Mobile Telecommunications-2000
Industrial Communications & Electronics, Inc., 13
industrial/scientific/medical (ISM) bands, 11
Industry Canada, 141
infrared technology, 86–87, 94
INMARSAT, 56, 59–61

INMARSAT-P, 66
Institute of Electrical and Electronics Engineers (IEEE), 19, 48–49
Integrated Digital Enhanced Network (iDEN), 80
intelligent transportation systems (ITS), 109
INTELSAT, 59–61
interaccess point protocol (IAPP), 50
International Association of Satellite Users and Suppliers, 141
International Mobile Telecommunications-2000 (IMT-2000), 45, 114, 128–129
International Telecommunications Union (ITU), 40, 45, 63–64, 111–115, 127–129, 141
Internet, 16, 90
  wireless access to, 87–88
intranets, *x*, 90
Iridium, 58, 62, 65–66
ISA cards, 34
ISM (industrial/scientific/medical) bands, 11
Italy, 23
ITS America, 142
ITS (intelligent transportation systems), 109
ITU, *see* International Telecommunications Union

Jacob Javits Convention Center, 44
Japan, 23, 115, 122

Kennard, William, on SBC-Ameritech merger, 118
Klever-Kart, 52
Kmart Corp., 44
Korea, 115
Kubrelecom, 37
Kyocera Corp., 122–124

LAAS (Local Area Augmentation System), 75
LATA (local access and transport area), 104
LECs (local exchange carriers), 33
LEOs, *see* low-earth orbit satellites
licensing, 3, 10–11
little LEOs, 57–59, 70–72
LMDS, *see* local multipoint distribution service
LNP (local number portability), 116
local access and transport area (LATA), 104
Local Area Augmentation System (LAAS), 75
local area networks, 1–2
local exchange carriers (LECs), 33
local multipoint distribution service (LMDS), 36, 86
local number portability (LNP), 116
location-tracking systems, 72–76
Lockheed Martin Corp., 21
Loral Orion Network Systems, 123
Loral Space and Communications Ltd., 67
low-earth orbit (LEO) satellites, 54, 57–59, 61, 64, 65, 67, 80
Lucent Technologies, 16, 35

MAC (medium-access-layer), 49
Major Trading Areas (MTAs), 10
manufacturing, 100
Matra Marconi Space S.A., 67
maximum permissible exposure (MPE), 25
McCaw, Craig, 13, 58, 64
medium-access-layer (MAC), 49
medium-earth orbit (MEO) satellites, 20, 70, 80
Metricom, 95
Microsoft, 92, 94
microwave communications, 80

MMITS, *see* Multimode Multifunction Information Transfer System
Mobile NC Reference Specification (MNCRS), 99
mobile satellite service (MSS), 19–20
mobile telephone switching office (MTSO), 4
Mobitex, 80
MOTIVATE, 126
Motorola, 3, 12, 13, 15–17, 57, 67, 80, 100, 123, 124
MPE (maximum permissible exposure), 25
MSS, *see* mobile satellite service
MTAs (Major Trading Areas), 10
Mtel, 15
MTSO (mobile telephone switching office), 4
multifunctional products, *ix–x*
multimedia applications, 123–124
Multimedia Telecommunications Association, 142
Multimode Multifunction Information Transfer System (MMITS), 131, 132

narrowband PCS, 10
National Cable Television Association, 142
National Care Information and Management Systems Society (HIMSS), 142
National Retail Federation, 143
National Telecommunications & Information Administration, 143
NCC (network control center), 71
NCs, *see* network computers
NEC, 93
network computers (NCs), 98–101
network control center (NCC), 71

Nextel Communications, Inc.,
12–14
Next-Generation Media
Technologies, 63
Nippon Telegraph & Telephone
Corp. (NTT), 125
Nokia, 85–86, 91, 92
NTT (Nippon Telegraph &
Telephone Corp.), 125

Odyssey, 59
OEMs (original equipment
manufacturers), 44
Omnibus Budget Reconciliation
Act of 1993, 10
OmniTRACS, 72–73
Orbital Communications Corp.
(ORBCOMM), 70
Orblink, 70
original equipment
manufacturers (OEMs), 44
Otis Elevator, 101

package-tracking systems, 81–82
pagers, 14, 80
answering machine, 16–17
digital voice, 16
and satellite communications,
53
two-way, 14–16
PalmPilot, 91, 92, 94
Panasonic, 16
PBXs, see private branch
exchanges
wireless, see wireless private
branch exchanges
PCI cards, 34
PCNs (personal communications
networks), 9
PCS, see personal communications services
PDAs, see personal digital
assistants
Personal Communications
Industry Association, 143

personal communications
networks (PCNs), 9
personal communications
services (PCS)
cellular communications vs., 7–9
deployment of, in United
States, 9–10
licensed vs. unlicensed, 10–11
personal computers, 1
personal digital assistants
(PDAs), 16, 47–48, 84–85
Personal HandyPhone (PHS), 23,
122
personal identification number
(PIN), 28
PHS, see Personal HandyPhone
PHY-specification, 49
PIN (personal identification
number), 28
"Pioneer's Preference," 10–11
Pittencrieff Communications,
Inc., 13
PocketNet, 85, 91
prepaid cellular calling service,
29
pricing
service, 22–24
for WLL, 39–40
zone-based, 39
private branch exchanges (PBXs),
10–11
wireless, 18
Progressive Insurance, 81
Project Angel, 38
PSTN (public-switched
telephone network), 62
public safety networks, 84
public-switched telephone
network (PSTN), 62

Qualcomm, Inc., 67, 72

radio frequency identification
(RFID), 20–22
RadioMall, 94

radios, software-definable,
130–132
Rand McNally, 10
Raytheon, 21
regional Bell companies, 38–39,
104–106
regulation, 27–28, 103–119
by FCC, 115–118
by Telecom Act, 103–110
trends in, 116–119
of universal service, 110–115
remote monitoring, 84
restaurants, 83
RFID, *see* radio frequency
identification
Ricochet, 86, 95
Rural Cellular Association, 143
Russian Federation, 37

SA (selective availability), 75
Satellite Broadcasting and
Communications Associa-
tion, 143
satellite communications, 19–20,
53–76
competition in, 63–64
costs of, 61–63
and location-tracking systems,
72–76
providers of, 64–72
trends in, 54–61
Savi Technologies, 21
SBC Communications, 118
Scientific Advisory Group on
Cellular Telephone Research,
143
Sear, Roebuck and Co., 82
Seiko Telecommunications, 17
selective availability (SA), 75
Sharp Corp., 17, 127
Siemens, 91
SIM cards, *see* subscriber identify
module cards
SkyBridge, 72
SkyTel, 14–15

smart cards, 30–31
SMR, *see* Specialized Mobile
Radio
Socket Communications, Inc.,
92–93
Sony Electronics, 122
Southwestern Bell, 39
Spaceway, 69–70
SPEAKeasy program, 131
Specialized Mobile Radio (SMR),
11–14, 134
SRI Consulting, 63
standards
"air interface," 6–7
future trends in, 127–130
for WLAN, 48–49
subscriber identify module (SIM)
cards, 30–31
Sun Microsystems, 124
Symbol Technologies, Inc., 21, 40

TCG (Teleport Communications
Group), 41
TDMA, *see* time division multiple
access
Telecommunications Act of
1996 (Telecom Act), 103–110,
117, 118
Telecommunications Industry
Association (TIA), 129, 143
teleconferencing, 126
Teledesic, 58, 64–65, 67
telepoint, 8–9
Teleport Communications Group
(TCG), 41
Texas Instruments, 17
thin-client/server computing,
96–98
third-generation (3G) mobile
telecommunications
technology, 45, 127–130
TIA, *see* Telecommunications
Industry Association
time division multiple access
(TDMA), 6–8

Timex, 17
Toshiba Corp., 72
TraansGuide, 21
TrueStream, 123
TRW, 59
II Morrow, Inc., 74
two-way messaging, 14–16

UMTS, *see* Universal Mobile
    Telecommunications
    System
underdeveloped countries, 36–37
Uniden Corp., 93–94
U-NII, 109–110
United Kingdom, 8–9, 37
United Parcel Service (UPS),
    81–82
Universal Mobile
    Telecommunications System
    (UMTS), 114–115, 127–128
universal service, 110–115
UPS, *see* United Parcel Service
U.S. Food and Drug Administra-
    tion (FDA), 26–27
U.S. GPS Industry Council, 143
U.S. Postal Service, 21
U.S. Telephone Association, 144
US West, 39
Utilities Telecommunications
    Council (UTC), 144

virtual displays, 126–127
voice pagers, 16
voice recognition, 125

WAAS (Wide Area Augmentation
    System), 75
WCS, *see* Wireless Communica-
    tions Service
Wide Area Augmentation System
    (WAAS), 75
wide area systems, 1–2
Windows CE, 92–94
WinStar Telecommunications,
    40–41, 109

Wireless Access, 14
Wireless Cable Association
    International, 144
Wireless Communications
    Service (WCS), 29–30
Wireless Data Forum, 145
Wireless LAN Interoperability
    Forum, 145
wireless local-area networks
    (WLANs), 18, 34–36, 46–52,
    80
  applications of, 50–52
  cost of, 48
  features of, 46–47
  standards for, 48–49
  trends in, 50
wireless local loop (WLL), 19,
    33–41
  justifications, 35–36
  pricing of, 39–40
  providers of, 40–41
  trends in, 37–38
  in underdeveloped economies,
    36–37
wireless private branch
    exchanges (WPBXs), 18–19,
    41–45
  future outlook for, 45
  justification for, 42–44
Wireless Telephone Protection
    Act, 28
WLANs, *see* wireless local area
    networks
WLL, *see* wireless local loop
workforce, mobility of, *ix*, 77, 89
World Health Organization, 27
World Radiocommunications
    Conference (WRC), 113–114
World Trade Organization
    (WTO), 115–116
World Wide Web (WWW), 90, 123
WPBX, *see* wireless private
    branch exchange
WRC (World Radiocommunica-
    tions Conference), 113–114

wristwatch communicators, 125

WTO, *see* World Trade
    Organization

WWW, *see* World Wide Web

zone-based pricing, 39